T0265357

Mapping State and Non-State Actors' Responses to

Nuclear Energy
in Southeast Asia

Mapping State and Non-State Actors' Responses to

Nuclear Energy in Southeast Asia

Editors

Nur Azha Putra
Energy Studies Institute, Singapore

Mely Caballero-Anthony
NTU, Singapore

World Scientific

NEW JERSEY · LONDON · SINGAPORE · BEIJING · SHANGHAI · HONG KONG · TAIPEI · CHENNAI · TOKYO

Published by

World Scientific Publishing Co. Pte. Ltd.

5 Toh Tuck Link, Singapore 596224

USA office: 27 Warren Street, Suite 401-402, Hackensack, NJ 07601

UK office: 57 Shelton Street, Covent Garden, London WC2H 9HE

British Library Cataloguing-in-Publication Data
A catalogue record for this book is available from the British Library.

**MAPPING STATE AND NON-STATE ACTORS' RESPONSES TO
NUCLEAR ENERGY IN SOUTHEAST ASIA**

ISBN 978-981-4723-19-0

Desk Editor: Sandhya Venkatesh

Typeset by Stallion Press
Email: enquiries@stallionpress.com

Printed in Singapore

CONTENTS

Introduction

NUCLEAR ENERGY: ISSUES AND CHALLENGES FOR SOUTHEAST ASIA STATES

Nur Azha Putra[1]

Introduction

Rising energy demands in Southeast Asia (SEA) will have to be met by more than just fossil fuels and renewable energy. As of 2011, primary energy demand was around 550 million tonnes of oil equivalent (Mtoe), which is 4.2 per cent of the total global demand.[2] The rising energy demand in SEA is driven mainly by increasing urbanisation and industrialisation. Additionally, a fifth or 133 million of the 600 million people in SEA still do not have access to electricity. Among the SEA states, Cambodia has the largest percentage share of energy poverty among its population with 66 per cent of its population has no access to electricity in 2011, followed by Myanmar (51 per cent), the Philippines (30 per cent), Indonesia

---

[1] Nur Azha Putra is Research Associate with the Energy Security Division at the Energy Studies Institute, National University of Singapore.
[2] 2013 World Energy Outlook: Southeast Asia Energy Outlook, p. 16. (http://www.iea.org/publications/freepublications/publication/southeastasiaenergyoutlook_weo2013specialreport.pdf: accessed 3 May 2013).

(27 per cent), Lao PDR (22 per cent), Vietnam (four per cent), and Malaysia and Thailand at one per cent each.[3]

With increasing rural electrification and rural-urban migration, the power generation needs in SEA are set to increase.[4] According to the International Energy Agency (IEA) estimates, the urbanisation rates in SEA are projected to increase from 45 per cent in 2011 to 59 per cent in 2035. This will lead to a significant increase in the total SEA urban population, which could reach more than 165 million by 2035. The largest increase in the urbanisation rate is set to take place in Malaysia at 82 per cent followed by Indonesia (66 per cent), the Philippines (59 per cent), and Thailand (45 per cent).[5]

Although IEA points out that increasing urbanisation will ultimately lead to a more productive and efficient use of energy, it will also inevitably lead to an overall net increase in energy demand and consumption especially in the residential and transport sectors. For instance, IEA observes a strong co-relation between urbanisation and rural electrification, and the growth in energy demand. As households become electrified and coupled with rising income, there will be an increase in demand for electronic services and goods such as cars, refrigerators, and air conditioners. Rising income is also set to increase at 3.7 per cent per year up to 2035, from $3,700 in 2011 to $8,700 in 2035.[6] As of 2011, the total primary energy demand in SEA totalled around 550 Mtoe.[7] Of the SEA states, Indonesia is the largest energy user. It consumes 36 per cent

---

[3] 2013 World Energy Outlook: Southeast Asia Energy Outlook, p. 26. (http://www. iea.org/publications/freepublications/publication/southeastasiaenergyoutlook_weo2013specialreport.pdf: accessed 3 May 2013).

[4] Ibid.

[5] 2013 World Energy Outlook: Southeast Asia Energy Outlook, p. 34. (http://www. iea.org/publications/freepublications/publication/southeastasiaenergyoutlook_weo2013specialreport.pdf: accessed 3 May 2013).

[6] 2013 World Energy Outlook: Southeast Asia Energy Outlook, p. 35. (http://www. iea.org/publications/freepublications/publication/southeastasiaenergyoutlook_weo2013specialreport.pdf: accessed 3 May 2013).

[7] 2013 World Energy Outlook: Southeast Asia Energy Outlook, p. 38. (http://www. iea.org/publications/freepublications/publication/southeastasiaenergyoutlook_weo2013specialreport.pdf: accessed 3 May 2013).

of the total energy demand in SEA followed by Thailand (21 per cent), Malaysia (13 per cent), and the Philippines (seven per cent). The remaining SEA states in total consumes as much as Thailand.[8]

Table 1: Southeast Asia: Primary energy demand by country (Mtoe).

	1990	2011	2020	2025	2035	2011–2035*
Indonesia	89	196	252	282	358	2.5%
Malaysia	21	74	86	106	128	2.3%
Philippines	29	40	58	69	92	3.5%
Thailand	42	118	151	168	206	2.3%
Rest of ASEAN	42	119	161	178	221	2.6%
Total ASEAN	223	549	718	804	1004	2.5%

Source: 2013 World Energy Outlook: Southeast Asia Energy Outlook, p. 38.
* Compound average annual growth rate.

Thailand is an energy importer. Its energy profile comprises oil (41.9 per cent), natural gas (26.7 per cent), coal (13.1 per cent), hydro (0.7 per cent), and others (mostly biomass 17.6 per cent).[9] In 2011, Thailand imported 45 per cent of its total energy needs — oil (68 per cent) and natural gas (28 per cent).[10] Natural gas and coal are the dominant fuels for power generation and industry. Diesel is the main fuel for the transport sector.[11] Thailand's economy is expected to grow at an annual rate of 3.5 to 4 per cent, according to the government.[12] This projection should be read with caution due

---

[8] 2013 World Energy Outlook: Southeast Asia Energy Outlook, p. 38. (http://www.iea.org/publications/freepublications/publication/southeastasiaenergyoutlook_weo2013specialreport.pdf: accessed 3 May 2013).

[9] The 3rd ASEAN Energy Outlook. (2011). p. 75. (http://www.energycommunity.org/documents/ThirdASEANEnergyOutlook.pdf: accessed 8 April 2014).

[10] Ibid.

[11] U.S. Energy Information Administration (2013). Thailand country profile. (http://www.eia.gov/countries/analysisbriefs/Thailand/thailand.pdf: accessed 21 March 2014).

[12] Ministry of Finance, Thailand: Thailand's Economic Outlook Projection 2013 and 2014, September 27, 2013. (http://www.mof.go.th/home/Press_release/News2013/102e.pdf: accessed 8 April 2014).

to the ongoing political uncertainty in the country. Thailand's population growth rate is less than one per cent per annum.[13]

Malaysia's national energy profile comprises mainly natural gas (63 per cent), coal (21 per cent), and hydro (10 per cent).[14] Its industries consumed about 60 per cent of the total energy used in the country.[15] The country's oil reserves are sufficient for another 33 years. However, the country has an abundance of natural gas and is the third largest global LNG exporter, after Qatar and Indonesia.[16] Malaysia's economic growth averaged 5.5 per cent since 2011 and is expected to continue growing at the same rate.[17] Malaysia is one of the several countries in SEA, which has recorded impressive economic growth in the last decade. Its economic growth is driven mainly by industry and in particular its manufacturing sectors. The country's average population growth since 2010 is 1.6 per cent per annum.[18]

Indonesia's energy mix comprises mainly oil (50 per cent), coal (22 per cent), natural gas (23 per cent), hydro (three per cent), and geothermal (two per cent).[19] Its energy demand is projected to grow at an average of 7 per cent per annum.[20] In 2011, its electricity demand stood at 40 Gigawatt — Electric (GWe) — and is expected to increase to 90 GWe in 2025 and 400 GWe in 2050.[21] The country's

---

[13] UNDATA: World Statistics Pocketbook (2014). (http://data.un.org/Country Profile.aspx?crName=THAILAND: accessed 17 April 2014).

[14] Malaysia Parliament records (January, 2013): Energy Security, Energy Efficiency, and Energy Dialogue. p. 6. (http://www.parlimen.gov.my/images/webuser/artikel/ro/halisah/Energy%20Security%20Halisah%20Ashari.pdf: accessed 4 August 2014).

[15] Ibid.

[16] Ibid.

[17] Malaysia: Economic Planning Unit (2014). (http://www.epu.gov.my/: accessed 4 August 2014).

[18] Department of Statistics: Malaysia @ a glance (2014). (http://www.statistics.gov.my/portal/index.php?option=com_content&view=article&id=472&Itemid=96&lang=en. accessed 8 August 2014).

[19] Energy Mix in Indonesia: A Review (2014). (http://seminar.pasca.uns.ac.id/download/enegrimix.pdf: accessed 4 August 2014).

[20] Ibid.

[21] Ibid.

oil reserves will be completely depleted by 2022.[22] However, Indonesia has an abundance of natural gas supplies. It is currently the second largest LNG exporter in the world, after Qatar.[23] Indonesia's economic growth averaged 6.3 per cent since 2011 and is expected to grow by the same average for the mid-term period.[24] Indonesia's population growth averaged 1 per cent per annum since 2010 and is set to continue at the same rate.[25]

The Philippines' energy mix in 2011 comprises oil (31 per cent), geothermal (22 per cent), coal (20 per cent), biomass (12 per cent), natural gas (eight per cent), hydro (six per cent), and biofuels (one per cent).[26] The largest energy demand is in the transportation sector followed by industrial, residential, and commercial sectors.[27] Total power generation in 2011 was 69 Gigawatt Hours (GWh).[28] The Philippines' economic growth has averaged 4.7 per cent between 2008 and 2012 and is projected to continue as much in the near future.[29] The country's population growth averaged 1.7 per cent between 2010 and 2015 and is unlikely to decline, at least in the next 10 years.[30]

Among the SEA states, Vietnam, Indonesia, Malaysia, Thailand, Philippines, and Myanmar have indicated their interest in the last

---

[22] Ibid.

[23] Ibid.

[24] Ibid.

[25] UNDATA: World Statistics Pocketbook (2014). (https://data.un.org/Country Profile.aspx?crName=Indonesia: accessed 17 April 2014).

[26] Philippine Department of Energy: Philippine Energy Plan (2012–2030). (https://www.doe.gov.ph/doe_files/pdf/Researchers_Downloable_Files/Energy Presentation/PEP_2012-2030_Presentation_(Sec_Petilla).pdf: accessed 1 August 2014).

[27] IEEJ Energy Policy Course: Philippines country presentation (2012). (http://eneken.ieej.or.jp/data/4482.pdf: accessed 3 August 2014).

[28] Philippine Department of Energy: Philippine Energy Plan (2012–2030). (https://www.doe.gov.ph/doe_files/pdf/Researchers_Downloable_Files/Energy Presentation/PEP_2012-2030_Presentation_(Sec_Petilla).pdf: accessed 1 August 2014).

[29] Philippines Country Profile: Asian Development Bank Outlook (2014). (http://www.adb.org/countries/philippines/economy: accessed 1 April 2014).

[30] UNDATA: World Statistics Pocketbook (2014). (https://data.un.org/Country Profile.aspx?crname=Philippines: accessed 17 April 2014).

decade to include nuclear energy power in their national fuel mixes. Vietnam has gone the distance and plans to construct two nuclear power plants (NPPs), with the support of Russia and Japan. However, due to safety and security issues, Vietnam announced in early 2014 that it would postpone the construction of its first NPP until 2020. Meanwhile, the remaining SEA states — Vietnam, Indonesia, Malaysia, Thailand, and the Philippines — have yet to decide on the use of nuclear energy. However, Indonesia remains the most likely and most prepared SEA nation after Vietnam to adopt nuclear energy. In November 2014, Indonesian media reported that national public acceptance on nuclear energy surged to 72 per cent after hovering around 60 per cent following Fukushima. Additionally, the government has issued a call for tender for the construction of a multi-purpose reactor (MPR) at its Serpong nuclear energy complex, which is situated near Jakarta. The construction of the MPR is scheduled to begin in 2016. Unlike Vietnam and Indonesia, it remains unlikely that Malaysia, Thailand, and the Philippines will be making any significant inroads towards its nuclear energy aspirations mainly on the basis that public acceptance in these states remain significantly lacking, among other things.

As for Singapore, the government decided in 2011 that the existing nuclear energy technology is unsuitable for the nation given the country's small size and population density.[31] Nevertheless, Singapore did not entirely dismiss the possibility of adopting nuclear energy in the future.[32] A theoretical study done by the Department of Civil Engineering at the National University of Singapore suggests that nuclear energy may be possible if the reactor is built underground.[33] A separate report suggests that the

---

[31] Ministry of Trade and Industry, Singapore: Factsheet: Nuclear Energy Pre-Feasibility Study (2011). (http://www.mti.gov.sg/NewsRoom/Documents/Pre-FS%20factsheet.pdf: accessed 5 August 2014).

[32] Singapore may get nuclear power plant, AsiaOne.com, 2 November 2010. (http://news.asiaone.com/News/AsiaOne+News/Singapore/Story/A1Story20101102-245235.html: accessed 5 August 2010).

[33] Andrew Palmer, Seeram Ramakrishna & Hassan Muzaffar Cheema (2010) *Nuclear power in Singapore*, The IES Journal Part A: Civil & Structural Engineering,

government considers the use of small modular reactors. However, as it stands, Singapore will not be adopting nuclear energy any time soon.

It is clear that the SEA states above are convinced that nuclear energy should be included in their national fuel mixes. Despite the huge financial outlay, nuclear energy generated electricity is price inelastic in the long run and is therefore resilient towards any economic and financial crises. The global financial crisis in 1997/1998 and the ensuing economic crisis in 2007/2008 have left a huge imprint on the SEA states. The era of cheap oil is gone and although SEA has entered the golden age of gas, fossil fuels alone are not enough to sustain their economic growth and future energy demands. Therefore, nuclear energy, despite the Fukushima incident, remains a viable long-term option even if it is not the primary fuel. At the current level of development, nuclear energy in SEA remains a matter of 'when' rather than 'if'.

Nuclear Energy in SEA: Issues and Challenges

The national energy strategies of the more economically developed SEA states — Vietnam, Thailand, Indonesia, Malaysia, and the Philippines — show that they are keen to include nuclear energy in their national fuel mixes in the future. Their successes however depend on two key policy challenges, public perception and nuclear energy safety and security.

Prior to Fukushima, there were already uncertainties among the SEA community with regard to the safety and security of nuclear energy. These uncertainties were largely shaped by past nuclear accidents such as 3-Mile Island in 1979 and Chernobyl in 1986. Despite those reservations, the SEA states remained largely confident and enthusiastic about their nuclear energy plans. However, the Fukushima incident in 2011 dented those aspirations resulting in all of the SEA states, with the exception of Vietnam, to declare that they needed more time to reassess their nuclear energy

3:1, 65–69, DOI: 10.1080/19373260903343449.

options. Understandably, public resentment towards nuclear energy was at an all-time high then. Public sentiments were not just shaped by the Fukushima incident itself but by the circumstances surrounding the Fukushima nuclear accident, such as the lack of information emanating from the Japanese authorities with regard to the accident. The situation was exacerbated by the fact that the incident happened in Japan where the Japanese's safety culture, has for years, been held up globally as one to be emulated. After all, the Japanese have been at the forefront of nuclear energy technology for many years and they have been manufacturing and managing their own nuclear power plants for decades. The Japanese have also been exporting their technological expertise and know-how to the SEA states. Before the Fukushima incident, the Japanese nuclear energy community was a picture of calm and measured reassurance and was a source of confidence to many nations. That image has since been shattered and it may take some time for the Japanese to recover.

Meanwhile, the public perception and confidence in SEA towards nuclear energy remains low. For instance, the Malaysian government has received stiff public denunciation from a coalition of protesters who are calling for the Australian company Lynas Corporation Ltd to halt the operations of its Lynas Advanced Materials Plant (LAMP), which is located in Kuantan, Malaysia.[34] LAMP, which is believed to be the largest rare earth refinery in the world, processes rare earth ore that is brought in from the mines in Western Australia.[35] Media reports claim that more than 1.2 million people have signed a petition calling for the plant to shut down its operations.[36] Protesters and green environmentalist groups are concerned that the plant, which stores radioactive elements that

---

[34] Arrests at Malaysian Rare Earths Refinery Protests (27 June 2014). *The Diplomat.com*, (http://thediplomat.com/2014/06/arrests-at-malaysian-rare-earths-refinery-protests/; accessed 5 August 2014).

[35] Ibid.

[36] Ibid.

can be spread through wind and water, may risk the health of the residents who live near LAMP. Media reports claim that there are approximately 30,000 people who live within two-kilometres of the factory, and a further 700,000 residents who live within 25 km.[37] Another major concern for the protesters is the fact that certain parts within Kuantan are vulnerable to heavy floods during the annual monsoon season.[38] Additionally, in May 2015, a nation-wide public survey shows that only two per cent of Malaysians have faith in the government's ability to manage nuclear energy facilities.[39]

While there are no official figures in Thailand, and the Philippines, public acceptance of nuclear energy in these countries appears to remain low as well. In 2013, Thailand's energy minister warned that the country is overly dependent on natural gas (70 per cent of Thailand's power generation comes from gas) and urged the government agencies to step up their efforts to gain public acceptance of nuclear energy.[40] As for the Philippines, in 2012, the government faced stiff opposition from civil society and threats from militant groups when a state official announced that it was considering reviving the country's nuclear energy plans.[41] The situation in Indonesia however is different.

Public acceptance in Indonesia has been steadily increasing despite the Fukushima incident and its troubled aftermath. A national poll taken in 2012 indicated that almost 53 per cent of the nation was in support of nuclear energy compared to 50 per cent shortly after the incident. In 2014, public acceptance increased to 72 per cent.

---

[37] Ibid.

[38] Nur Azha Putra, Hunger for energy and pull of N-power (23 April 2014), *The Straits Times.*

[39] Only 2% Malaysians trust our govt with nuclear technology. (2015). *Astro Awani News,* (http://english.astroawani.com/malaysia-news/only-2-malaysians-trust-our-govt-nuclear-technology-64399; accessed 21 January 2016).

[40] Ibid.

[41] Ibid.

As ASEAN grows closer as an integrated regional community, the issue of nuclear energy safety and security can no longer be viewed as solely in the domain interests of the individual member states. Although the issue of safety and security is first and foremost the responsibility of each member state, the effect of a nuclear energy disaster will have repercussions that go beyond national borders. This is even more pressing considering the close proximity of the SEA states. It is therefore in the collective interest that the SEA states develop a regional nuclear energy emergency preparedness plan, which cannot be achieved in the absence of certain factors.

First, there must be a regular regional Track One forum to discuss the issue of nuclear energy safety and security. Over the years, ASEAN has established several such networks, which include the ASEAN Nuclear Energy Cooperation Sub-Sector Network (NEC-SSN)[42] and the ASEAN Network of Regulatory Bodies on Atomic Energy (ASEANTOM).[43] Both networks bring together senior officials from the ASEAN member states to exchange information and work together towards regional cooperation on nuclear energy regulation. However, information and issues that are discussed at both networks are not readily available to the public, which thus implies that these networks could strive for greater transparency.

Secondly, Track One initiatives should be complemented with Track Two initiatives as well. Unlike Track One networks, which typically comprise senior government officials only, Track Two networks activities include members who are from the larger policy-making community that includes academics, social scientists, researchers and ex-government officials. One of the implications of the Track Two network activities is that it helps to expand the policy making community to include non-government officials

---

[42] ASEAN Centre for Energy: NEC-SSN. (http://aseanenergy.org/index.php/acebodies/nec-ssn; accessed 15 August 2014).

[43] ASEANTOM. (2015). (http://aseantom.blogspot.sg/; accessed 28 December 2015).

who are able to provide expert knowledge and insights on nuclear energy to decision-makers. Perhaps more importantly, Track Two networks could bridge the knowledge and information gap between states and within states. After all, nuclear energy and its related safety and security issues are relatively new to the ASEAN region and the region is still trying to develop and expand its knowledge base. Examples of Track Two networks that have done some work on nuclear energy safety and security include the Economic Research Institute for ASEAN and East Asia (ERIA)'s working groups on 'International Cooperation on Nuclear Safety Management in East Asian countries'[44] and the Council for Security Cooperation in Asia-Pacific (CSCAP)'s Nuclear Energy Experts Group.[45] On that note, this volume seeks to raise awareness at the policy level of the potential nuclear energy policy issues on safety and security that are relevant to ASEAN.

In Chapter 1, Alistair Cook and Sofia Jamal analyse the policy trends and processes among the Southeast Asian states and use it to identify the signposts, which could be used to craft a regional energy security policy in the future. In Chapter 2, Nicholas Fang reviews the existing state of multilateral cooperation on nuclear issues in the ASEAN region, assessing the need for a more formal regional nuclear safety regime, and analyses the future prospects for ASEAN in this area. In Chapter 3, Eulalie Han looks at the overall nuclear energy safety and security discourse in Southeast Asia and identifies several potential areas for nuclear energy cooperation among the ASEAN member states. In Chapter 4, Aisha Bidin examines the issues and challenges for Malaysia's legal framework *vis-à-vis* the country's nuclear energy aspirations and in the broader scheme of things, how the country's legal framework

---

[44]Murakami, Tomoko (2013). Study on International Cooperation Concerning Nuclear Safety Management in East Asian Countries, ERIA Research Project Report 2012-28, June 2013, (http://www.eria.org/publications/research_project_reports/FY2012-no.28.html; accessed 10 August 2014).

[45]CSCAP: Nuclear Energy Experts Group. (http://www.cscap.org/index.php?page=nuclear-energy-experts-group-neeg: accessed 15 August 2014).

could adhere to the international legal instruments. In Chapter 5, John Bauly offers his perspectives on the various aspects of nuclear policy and governance that Singapore could adopt. In the penultimate chapter, Chapter 6, Ton-nu-thi Ninh provides a brief overview of Vietnam's nuclear energy plans and identifies several policies, governance, and human resource issues that the country faces. In the final chapter, Chapter 7, Hans Roegner provides a crucial perspective and critical insights on the positive and negative trade-offs and the socioeconomic impact of nuclear energy for new nuclear energy countries.

Chapter 1

NUCLEAR ENERGY SECURITY AND THE POLICY ENVIRONMENT IN SOUTHEAST ASIA

Alistair D.B. Cook and Sofiah Jamil[1]

Introduction

Energy demand in Southeast Asia (SEA) is rising in tandem with economic development which poses challenges to the energy security of states in the region. Indonesia, Malaysia, Singapore, the Philippines, Thailand and Vietnam account for 95 per cent of the region's energy consumption. Available national supply in these countries however is not sufficient to meet demand. Although Indonesia is a major exporter of coal and liquid natural gas, it is a net importer of oil; Malaysia exports natural gas but imports coal; Singapore and the Philippines are dependent on imports. A major source of energy insecurity in the region is the high dependence on fossil fuels, which make up 73 per cent of the region's energy

---

[1] Alistair D. B. Cook is a Research Fellow and Sofiah Jamil is an Adjunct Research Associate at the RSIS Centre for Non-Traditional Security Studies at the S. Rajaratnam School of International Studies, Nanyang Technological University, Singapore. The authors would like to thank Lina Gong for her research assistance with the presentation upon which this chapter is based.

consumption.[2] A high proportion of these fuels have to be imported from outside of SEA, which means that the region is susceptible to risks related to geopolitics and domestic insecurity in oil-exporting countries and regions illustrating the complexity of energy security.

Adding to this complexity in SEA is the diversity of the region in terms of each country's energy needs as well as varying levels of availability and access to these energy resources. Natural resources are usually situated far away from population centres which are focused on coastal areas, disproportionately affected by climate change and also separated by water. Indonesia, the largest energy consumer in the region with 36 per cent of overall energy demand, consumes 66 per cent more than Thailand, the region's second largest user. Indonesia exports steam coal and liquefied natural gas, whereas Thailand is heavily dependent on energy imports. Vietnam is in transition from its own natural resources of fossil fuels and renewables to becoming an energy importer. There is varied access from Singapore, Brunei, Thailand, Malaysia, which have near universal access to less than 50 per cent access in Cambodia and Myanmar.[3] It is these diverse concerns that shape energy policy and the energy mix of countries in the region. Our chapter analyses the policy trends and processes in SEA to offer signposts on the way ahead in energy security policy.

With diverse national concerns but a general dependence on fossil fuels in the region, nuclear technology is seen as an attractive addition to develop a more varied energy mix. Vietnam plans to build ten Nuclear Power Plants (NPPs) by 2030; Indonesia plans to build four nuclear reactors by 2024; and Thailand intended to build five NPPs by 2030 but these plans were suspended. Malaysia and the

---

[2] Jamil, Sofiah and Lina Gong (2013) Nuclear energy development in Southeast Asia: Implications for Singapore, NTS Insight, no. IN13-01 Singapore: RSIS Centre for Non-Traditional Security (NTS) Studies (Nuclear energy development in Southeast Asia: Implications for Singapore; http://www3.ntu.edu.sg/rsis/nts/HTML-newsletter/insight/INTS-insight-mar-1301.html, accessed 14 January 2015).
[3] IEA/ERIA (2013), Southeast Asia Energy Outlook, Paris: OECD/IEA (https://www.iea.org/publications/freepublications/publication/SoutheastAsiaEnergyOutlook_WEO2013SpecialReport.pdf, accessed 31 January 2015).

Philippines on the other hand are studying the option and Singapore commissioned a pre-feasibility study and subsequently decided against the nuclear energy option.[4] The Fukushima crisis slowed down the momentum that had built up in the region as the crisis led to strong opposition to NPPs due to safety concerns. The crisis, however, has not completely stopped the nuclear energy programmes and countries have now taken extra effort to address safety issues related to NPPs. For example, Vietnam and Indonesia have tried to persuade communities through socialisation initiatives. Vietnam arranged for heads of villages close to potential sites of nuclear power plants to visit plants in Japan, including Fukushima in 2010. Such initiatives appear to have paid off and there is high support for nuclear energy projects in the country.[5] Given the susceptibility of SEA region to natural disasters, NPPs in the region are a cause for concern for national leaders and governments. Indeed, the absence of a nuclear safety protocol in ASEAN further heightens their concern.

At present, the only agreement on nuclear safety that ASEAN member countries have signed and implemented is the 1995 Southeast Asian Nuclear-Weapon-Free Zone (SEANWFZ) Treaty, which came into force in 2001. The treaty is a nuclear weapons moratorium between the ten ASEAN member states. In the treaty, Article 4 states that there will be no prejudice toward the peaceful use of nuclear energy. The treaty also enshrines in law that prior to embarking on civilian nuclear energy programs political buy-in is needed from the International Atomic Energy Agency (IAEA) and from other ASEAN member states. Indeed, nuclear non-proliferation concerns and safeguards will be very important as ASEAN proceeds in developing its nuclear power capabilities. Under IAEA guidelines, when a safeguards agreement enters into force, a state

---

[4]Ministry of Trade and Industry (2012). Singapore Pre-feasibility Study on Nuclear Energy, Singapore: NERA Economic Consulting (https://www.mti.gov.sg/NewsRoom/Documents/Pre-FS%20factsheet.pdf; accessed 31 January 2015).
[5]Jamil, Sofiah and Lina Gong (2013). Nuclear energy development in Southeast Asia: Implications for Singapore. *INSIGHT, a publication of the Centre for Non-Traditional Security (NTS) Studies*; accessed 14 January 2015.

has an obligation to declare to the IAEA all nuclear material and facilities subject to safeguards under the agreement. In 2010, ASEAN agreed that the Nuclear Energy Cooperation Sub Sector Network (NEC-SSN) would serve as the key body to assist the ASEAN members in their civilian nuclear energy cooperation but there has been little progress since then. In contrast to the limited activity at the regional level in SEA, in Europe the European Atomic Energy Community (EURATOM) provides the guarantee over safe nuclear energy development. This development provides a key reference point for the construction of a regional response in SEA.

The NEC-SSN is a regional forum for the ASEAN member states to engage on the development of safe civil nuclear energy. Since its inception in 2007, there has been limited progress and interest on developing an intra-ASEAN agreement. However, the ASEAN +3 Forum on Nuclear Energy Safety met in 2009 "to enhance synergy on the peaceful uses of nuclear energy in the region, particularly in terms of technology transfer and capacity — building", illustrating the wider implications and number of states interested in the development of civilian nuclear power in SEA. More recently in 2011, the East Asia Summit participants encouraged regional compliance with United Nations' non-proliferation commitments and welcomed the conclusion of negotiations on the SEANWFZ Protocol, which will see the five nuclear powers sign on to the regional nuclear weapons free zone. ASEAN +3 and East Asia Summit Energy Ministers also meet annually, with nuclear energy on the agenda for both groupings.[6]

However, in the absence of a regional agreement on the use of civil nuclear power, there are many questions facing the SEA

---

[6] The James Martin Center for Nonproliferation Studies (CNS, Monterey, United States), the Center for Energy and Security Studies (CENESS, Moscow, Russia) and the Vienna Center for Disarmament and Non-Proliferation (VCDNP, Vienna, Austria) (2012). Identifying Mechanisms and Approaches to Address Nuclear Security in Southeast Asia. In *Prospects for Nuclear Security Partnership in Southeast Asia*, Monterey/Moscow/Vienna: CNS, CENESS, VCDNP.

governments. These concerns cover the economic impact of nuclear power development, the environmental impact and safety of nuclear power plants, and the security implications of civilian use of nuclear power in the region. Firstly, there is a trade-off for governments to consider, which on the one-hand is the low cost of nuclear power compared to other forms of power generation, but with a high capital investment compared to other power generation such as gas. Secondly, the vast majority of nuclear energy is generated from enriched uranium and at present there are no known uranium reserves in SEA thus making states in the region susceptible to external constraints. While there is the possibility to use recycled uranium from weapons, which offers more source diversity, it is in diminishing supply.[7] Thirdly, as SEA is a large earthquake zone, there are significant concerns from states across the region about where nuclear power plants are located and the risks posed. Fourthly, there is concern that nuclear material could be used to develop Weapons of Mass Destruction by non-state actors and threaten peace and security in the region particularly in states with a weak governance record. However, these concerns have not significantly impacted the development of the civilian use of nuclear power in SEA and the wider East Asian region.

Before the Fukushima crisis, nuclear power accounted for 30 per cent of Japan's energy mix in 2006, and it was predicted that it would reach 40 per cent by 2030. However, in the aftermath of the Fukushima disaster, the Japanese government created three organisations, the Council on Energy and Environment, Nuclear Regulation Agency (NRA), and the Fundamental Policies Committee to review the country's nuclear energy commitment. The Japanese government decided that they would phase-out nuclear energy by 2040. However, the change in government in Japan saw the redaction of its nuclear-free ambition. By the end of

---

[7]Symon, A. (2008). Nuclear Power in Southeast Asia: Implications for Australia and Non — Proliferation, Analysis, Sydney: Lowy Institute for International Policy (http://www.lowyinstitute.org/files/pubfiles/symon-nuclear-power-in-southeast-Asia.pdf; accessed 14 January 2015).

2013, a revision of its nuclear energy policy was carried out by the industry ministry's advisory committee for natural resources and energy subcommittee. The new Japanese policy was that there would be no government plan to include a fixed goal for reliance on nuclear power due to new safety standards. The impact and development of public opposition or trust will shape Japanese nuclear energy policy and determine whether nuclear power is needed for a sustainable energy mix.[8] However, recent developments and longer term energy needs point to a nuclear energy restart.

In SEA, there are varied responses in the wake of the Fukushima incident and the development of nuclear energy security policy. Refer to Table 1 for a summary of the impact of the Fukhushima nuclear accident on the Southeast Asian states. Prior to the incident, Vietnam and Indonesia both developed plans for nuclear power plants, and the Fukushima incident did not significantly impact their policy goal of developing civilian use of nuclear power. Thailand also developed a plan to build nuclear power plants but in the aftermath of the Fukushima incident, chose to review its safety record. It has subsequently decided to postpone its decision for three years on two separate occasions. These two responses reflect the mixed initial responses to Fukushima reflecting the different domestic political situations but are also responses that illustrate a commitment to civil nuclear power in the long-term. There are however, common energy themes such as a shift towards the salience of energy security as a security concern, and reducing economic costs as a result of imports in a period of high global energy prices. There are also emerging themes to improve sustainability driven by concerns over what the local population want, particularly as the region most vulnerable to climate change.

---

[8]World Nuclear Association (2013). Nuclear Energy in Japan, London: WNA. (http://www.world-nuclear.org; accessed 10 January 2015).

Table 1: Impacts of Fukushima on nuclear plans in East Asia.

Country name	Share in electricity generation mix before 2011	Five year pass	Pass suspended	ERIA 2013 assessment	Projection mode after 2011
Vietnam	×	√	×	√	10.1% (2030)
Indonesia	×	√	×	√	4% (2025)
Thailand	×	√	√	×	8% (2030)
Malaysia	×	√	√	×	6% (2030)
Singapore	×	√	×	×	×
The Philippines	×	×	×	×	×
China	1–2%	×	√	√	4% (2020)
Japan	30%	√	√	√	×
South Korea	30%	√	×	√	40% (2030)
India	2%	√	×	√	25% (2050)

Source: Compiled by Lina Gong, RSIS Centre for Non-Traditional Security Studies, Singapore, October 2013.

Drivers of Nuclear Energy Plans in Southeast Asia

While international efforts to cooperate on nuclear safety is com-
mendable, there is a need to examine what drives or motivates
governments to pursue the nuclear safety agenda? At the crux of the
matter, states are pursuing civil use nuclear energy to support their
economics needs. There are several factors influencing this trend.

Meeting Economic Needs

Firstly, Southeast Asian states efforts to meet their energy needs are
made easier with the available supply of nuclear technology/
expertise. The supply particularly comes from countries that are
willing to export their nuclear capabilities, such as Russia and
Northeast Asian countries, as a means of strengthening their own
economies and soft power in the region. There are several ways in
which this is pursued. One of the earliest forms has been to provide
nuclear technology and expertise in the form of overseas development assistance (ODA). The Korea International Cooperation
Agency (KOICA), for instance, has been providing aid to Southeast
Asian countries since the 1990s for the development of civil nuclear
energy plans.[9] The significance of ODA is clear in Vietnam, where
the capital for Vietnam's first proposed nuclear power plant in
Ninh Thuan is entirely dependent on ODA from Russia and
Japan.[10] Post-Fukushima Japan under the Abe government has also

---

[9] For details on South Korea's overseas development assistance (ODA) efforts
through its Korean International Cooperation Agency (KOICA), see: He, Lisa
(2011). South Korea's nuclear development assistance in Southeast Asia: The
implications and challenges of the security environment in the 21st century,
Washington, DC: Joint US-Korea Academic Studies, (http://keia.org/sites/
default/files/publications/emergingvoices_final_lisahe.pdf; accessed 10
November 2014).

[10] Ninh Thuan nuclear power plants may not be kicked off as schedule
(27 September 2012). *Vietnam.Net Bridge*, (http://english.vietnamnet.vn/fms/
special-reports/50577/ninh-thuan-nuclear-power-plants-may-not-be-kicked-off-
as-schedule.html; accessed 10 November 2014).

sought to export nuclear expertise to revive its stagnating economy, such as working with France to build Turkey's second nuclear power plant.[11]

Aside from nuclear technology and expertise, SEA can potentially tap on to uranium supply within the Asia-Pacific region. Australia, which is the third largest exporter of uranium after Canada and Kazakhstan,[12] could be a potential source for nuclear energy programmes in SEA. Australia's uranium exports made up about 33 per cent of Australia's energy exports[13] and is currently supplying uranium to Northeast Asian countries[14] and sealing its deal with India.[15] SEA would thus be a potential market for Australia in the future.

Another factor driving the nuclear energy agenda in SEA is the high costs of other alternative sources of energy. Despite the great potential of utilising geothermal energy in Indonesia, the government has noted the lack of investments to develop such projects. To date, the Indonesian government has only been able to allocate USD 173.5 million from a total budget of about USD 1.3 billion to finance geo-thermal energy development.[16] In this regard, several energy scholars have noted that nuclear energy

---

[11] Slodkowski, Antoni, France, Japan join forces for larger share of nuclear market (7 June 2013). *Reuters*, (http://www.reuters.com/article/2013/06/07/us-japan-france-nuclear-idUSBRE9560C720130607; accessed 10 November 2014).

[12] World Nuclear Association (2015). Australia's Uranium, (http://www.world-nuclear.org/info/Country-Profiles/Countries-A-F/Australia/#.UaHAPZxji5w; accessed 10 January 2015).

[13] Bureau of Resources and Energy Economics, (2012). Energy in Australia 2012, Department of Resources, Energy and Tourism, (http://www.world-nuclear.org/info/Country-Profiles/Countries-A-F/Australia/#.UaHAPZxji5w; accessed 20 December 2014).

[14] World Nuclear Association (2015), op.cit; accessed 10 January 2015.

[15] Edwards, Michael, Uranium talks to dominate Gillard's India visit (16 October 2012). *ABC News*, (http://www.abc.net.au/news/2012-10-16/uranium-talks-to-dominate-gillards-india-visit/4314824; accessed 20 November 2014).

[16] Indonesia mulls further developing geothermal power: minister (6 October 2013). *Xinhua* (http://www.sseabp.com/Item/8772.aspx; accessed 20 November 2014).

would be a "mid-term" option until alternative energy sources become more affordable.

Managing Stakeholders

Based on these economic needs, nuclear development plans in SEA are contingent on the governments' ability to manage stakeholders from various sectors and levels. Increasing public awareness through better communication and engagement of various sectors is a crucial aspect, but has not always produced successful results. Socialisation programmes in Vietnam, for instance, have been fairly successful, where Vietnamese communities living near proposed nuclear reactor sites are sent to Japan to visit communities living near nuclear power plants, as a means of allaying any fears toward nuclear energy projects.[17] One could however argue that there is an element of propaganda given the tight control over the information that the villagers received. Conversely, socialisation programmes in Indonesia have not been as successful. For instance, while the national government had allocated USD15.9 million to increase public awareness in the Bangka-Belitung (Babel) province, local communities were 'unhappy with socialisation programmes taking place in their villages, and also remained unconvinced that Bangka was geologically safe for a nuclear power plant'.[18]

Aside from galvanising the public to accept nuclear energy projects, some governments may choose to silence oppositional voices to further the interests of the government and other lobby groups. This is perhaps easier to do in countries with centralised political systems, such as Vietnam. Countries with democratic

---

[17] Stour, Kimura (2012). Japanese nuclear power generation comes to a Vietnamese village'. The Asia-Pacific Journal, 10 (37) No.3 (http://www.japanfocus.org/-Kimura-Satoru/3824; accessed 20 December 2014).

[18] Fauzan, A.U. and Schiller, J. (2011). After Fukushima: The rise of resistance to nuclear energy in Indonesia, German Asia Foundation/Asia House, Essen, No. 23, p. 23–24, (http://www.asienhaus.de/public/archiv/resistance-in-indonesia-after-fukushima.pdf; accessed 20 December 2014).

political systems and freedom of media have had more difficulties in silencing opposition, as seen in Malaysia[19] and Thailand, while the opposition groups in Indonesia[20] include the religious leaders.

Mob-Mentality on Nuclear Safety?

Another driver of nuclear energy agendas is the growing multilateral efforts in enhancing Nuclear Safety to mitigate accidents, advance technological levels for better safety standards, and improve governance capabilities.[21] Nuclear safety has been part of several international meeting agendas including the 2012 Asia Pacific Economic Cooperation (APEC) meeting's St Petersburg declaration[22] and ASEAN, ASEAN Plus Three and the East Asian Summit frameworks.

That said, however, there may be some degree of jumping on the bandwagon. While there is much use of the nuclear safety rhetoric, how much of this rhetoric translates into action? In this regard, it is still unclear to what extent regional and national mechanisms are able to be effective in mitigating the potential challenges stemming from nuclear energy development. The next section will demonstrate how the inability to address gaps in the nuclear life cycle will pose challenges to governments and communities.

---

[19] Calls renew for Lynas shutdown after third death at plant (14 December 2013), *The Malaysian Insider*, (http://www.themalaysianinsider.com/malaysia/article/calls-renew-for-lynas-shutdown-after-third-death-at-plant; accessed 10 December 2014).

[20] APSNet Policy Forum (2007). Nuclear fatwa: Islamic jurisprudence and the Muria nuclear power station proposal, (http://nautilus.org/apsnet/nuclear-fatwa-islamic-jurisprudence-and-the-muria-nuclear-power-station-proposal/; accessed 10 December 2014).

[21] See Asian Nuclear Safety Network (http://ansn.iaea.org/default.aspx: accessed 5 December 2014).

[22] APEC (2012). St. Petersburg Declaration — Energy Security: Challenges and Strategic Choices. *2012 APEC Energy Ministerial Meeting*, Saint Petersburg, Russia, 24–25 June 2012, (http://apec.org/Meeting-Papers/Ministerial-Statements/Energy/2012_energy.aspx; accessed 5 December 2014).

Challenges: Gaps in Nuclear Cycle Undermine Development

The lack of attention to the various stages of the nuclear energy development cycle — from construction, generation to waste disposal — can result in ineffective management of nuclear energy. This is particularly so when there is limited consideration of the environmental risks in nuclear energy development as part of the trade and bilateral growth. This includes limited consideration of high risks with low-probability[23] (e.g. tsunami and earthquake in Fukushima) to save costs, as well as medium-impact risks that accumulate overtime such as the impacts on health and food security. In terms of the former, several studies[24] have discovered links between lung cancer and uranium mining. This therefore highlights non-communicable diseases, which are on the rise but have not had as much attention as communicable diseases in human development.[25] In terms of the latter, improper nuclear waste disposal will have impacts on food security too since NPPs

---

[23] Schweitzer, J., 'Nuclear Energy Survives Only on the Basis of Faulty Risk Assessment (9 September 2013). *Huffington Post*, (http://www.huffingtonpost.com/jeff-schweitzer/nuclear-energy-survives-o_b_3882434.html; accessed?). Lee, B., Preston, F., Green, G. (2012). Preparing for high-impact, low-probability events: lessons from Eyjafjallajökull — A Chatham House Report, The Royal Institute of International Affairs, January 2012, (https://www.chathamhouse.org/sites/files/chathamhouse/public/Research/Energy,%20Environment%20and%20Development/r0112_highimpact.pdf; accessed 5 January 2015).

[24] Mulloy, K.B., James, D.S., Mohs, K and Kornfeld, M. (2011) Lung cancer in a nonsmoking underground uranium miner. *Environmental Health Perspectives*, 109(3), (http://www.ncbi.nlm.nih.gov/pmc/articles/PMC1240251/; accessed 5 December 2014); Finkelstein, M.M. (1996). Clinical measures, smoking, radon exposure, and risk of lung cancer in uranium miners, *Occupational and Environmental Medicine*, no 53, pp. 697–702.

[25] Friel, S., Bowen, K., Campbell-Lendrum, D., Frumkin, H., McMichael, A.J., and Rasanathan, K. (2011). Climate Change, Noncommunicable Diseases, and Development: The Relationships and Common Policy Opportunities, *Annual Review of Public Health*, Vol. 32, pp. 133–147. (http://www.who.int/sdhconference/resources/friel_annualrevpubhealth2010.pdf; accessed 15 January 2015).

are typically located near the coasts, nuclear leaks at sea can affect the fisheries sector, which contributes substantially to several ASEAN economies. Indonesia, for example, is the largest producer of fishery products in SEA and fisheries contributes to 19.2 per cent of GDP in the country's agricultural sector.[26] The consumption of contaminated fishery products will have adverse health implications.

Conclusion

Moving forward, there are four aspects that need to be strengthened and sensitised to a Non-Traditional Security approach.[27] Firstly, in addressing future challenges, proposed solutions must comprehensively review current global standards of human security in light of the implications of nuclear energy development as mentioned above. Primary standards as seen in the Millennium Development Goals (MDGs) and the Sustainable Development Goals (SDGs) should only serve as a baseline and there is therefore a need to raise the bar to improve the quality of health, food and environmental security in the long term.

Secondly, efforts must be taken to ensure clear accountability. These include building public confidence via regular updates of nuclear-related information, and more transparent engagement and communication with various members of civil society and academic community (including understanding factors that limit their involvement in policy advising and making). This is vital for optimal nuclear safety standards as it needs greater transparency to

---

[26] Directorate for Agribusiness Planning (2011). Fisheries Industry at a Glance, *Agribusiness Update*, Department of Investment Planning, p. 1, (http://www3.bkpm.go.id/img/file/fisheries.pdf; accessed 15 January 2015).

[27] An NTS approach/framework incorporates the following characteristics: participation, fair-and-accessible legal framework, transparency, responsiveness, consensus-orientated decision making process, equity and inclusiveness, effectiveness and efficiency, accountability. For details, please see Mely Caballero-Anthony, Alistair D. B. Cook, (eds.) (2013). *Non-Traditional Security in Asia: Issues, Challenges and Framework for Action*, Singapore: ISEAS Publishing.

overcome potential leaks, facilitate decommissioning processes and better nuclear waste disposal.

This brings us to the third aspect of transparent nuclear policies. While democratic process is important in ensuring accountability and transparency, it can also undermine nuclear policies in the event of a change in government. In this regard, one could argue that Vietnam's long-term nuclear energy plans could possibly be more stable given its centralised one-party system as opposed to the risks of changing policies by different political parties in a democratic system.

Finally, multilateral collaboration in operationalising proactive rather than reactive policies is necessary to effectively monitor nuclear energy development in the region. This would include strengthening ASEAN +3 and EAS arrangements on improving safety standards and energy markets, and advance disaster management/preparedness efforts so as to address root causes of disasters, and not simply its symptoms.

Chapter 2

TOWARDS A REGIONAL NUCLEAR ENERGY SAFETY REGIME IN SOUTHEAST ASIA

Nicholas Fang[1]

Introduction

Before the March 2011 disaster at Fukushima Daiichi, Southeast Asian countries were rapidly moving towards the adoption of nuclear power as part of their national energy mix. Prior to the accident, Thailand, Malaysia, and Indonesia all planned to open their first nuclear power plants by 2020, while Vietnam aimed to have its first plant operational as early as 2015. In the aftermath of the Fukushima incident, Thailand, Malaysia and Indonesia appear to be reassessing their plans, and no new firm timelines for nuclear energy development have been announced by them. However, none of the three countries have ruled out nuclear energy entirely, and the current pause seems to represent a delay, rather than a complete abandonment of plans (Tay and Phir, 1990).[2]

---

[1] Executive Director, Singapore Institute of International Affairs.
[2] Tay, Simon S.C. and Phir Paungmalit (2012). Critical environmental questions: nuclear energy and human security in Asia. In *Nuclear Power and Energy Security in Asia*. Rajesh Basrur and Koh Swee Lean Collin (eds.), pp. 90–111, New York; Abingdon: Routledge.

In contrast to its neighbours, Vietnam has made it clear that it will definitely be proceeding with the construction of nuclear power plants, albeit on a more cautious timescale. Vietnam was scheduled to start building its first nuclear plant in 2014, but construction has now been delayed until 2020. In theory, this is to allow more time to raise the country's safety standards, though some observers say the delay has more to do with negotiations on financing and technical matters. Regardless, this is only a momentary setback, and it is evident that Vietnam remains serious about nuclear power development. Vietnam already has signed contracts with Russian and Japanese companies to supply the country's first nuclear reactors (KAS and EU Centre in Singapore 2012).[3] In June 2013, Vietnam initiated a joint feasibility study with South Korea, which could lead to a similar deal being signed. The US may also soon start selling nuclear fuel and technology to Vietnam; in March 2014, United States President Barack Obama approved a bilateral nuclear deal between the US and Vietnam that is now awaiting Congressional approval. Once Vietnam begins building or operating its nuclear power plants, it is possible other countries in the Association of South-east Asian Nations (ASEAN) could be motivated to revisit their own plans (Fang and Choo, 2014).[4]

Despite the prospect of operational nuclear power plants in ASEAN's neighbourhood within the next two decades, regional cooperation on nuclear safety is still in its infancy. This paper reviews the current state of multilateral cooperation on nuclear issues in ASEAN, assesses the need for a more formal regional nuclear safety regime, and examines the future prospects for ASEAN in this area.

---

[3] Konrad Adenauer Stiftung and the EU Centre in Singapore (2012). Workshop on Nuclear Energy and Nuclear Safety, p. 9, Singapore: KAS.

[4] Fang, Nicholas and Choo, Aaron. ASEAN must learn from the Fukushima disaster (14 March 2014), *TODAY*. (http://www.todayonline.com/commentary/asean-must-learn-fukushima-disaster; accessed 17 March 2014).

Where is ASEAN Now?

In principle, ASEAN countries have affirmed their commitment to regional cooperation on nuclear energy safety. At the 13th ASEAN Summit in November 2007, ASEAN leaders signed the ASEAN Declaration on Environmental Sustainability, pledging "to forge ASEAN-wide cooperation to establish a regional nuclear safety regime", building on the principles established in past ASEAN commitments such as the Southeast Asia Nuclear-Weapon-Free Zone (SEANWFZ) Treaty.

SEANWFZ was originally signed in December 1995, entering into force in March 1997. SEANWFZ is a non-proliferation treaty, and thus relatively little of the agreement directly pertains to nuclear energy. The SEANWFZ treaty does require countries to report any "significant event" regarding their nuclear status. However, what constitutes such an event is ill-defined, and no ASEAN country has ever made such a report, even when setting up research reactors and establishing plans to build civilian nuclear power plants. That said, SEANWFZ does in principle oblige countries to subject national peaceful nuclear energy programmes to safety assessments according to International Atomic Energy Agency (IAEA) standards, and make those safety assessments available to other state parties upon request. The more recent Plan of Action to Strengthen the Implementation of the SEANWFZ Treaty (2007–2012) also encouraged ASEAN member states to join or complete accession to IAEA agreements and work towards a regional nuclear safety regime.

ASEAN officially established a Nuclear Energy Cooperation Sub-Sector Network (NEC-SSC) in July 2010. The NEC-SSN is working to develop an ASEAN Action Plan on Public Education on Nuclear Energy and Nuclear as the Clean Energy Alternative Option, "with a view to enhancing public awareness and acceptance of the usage of nuclear energy for power generation" (Joint Ministerial Statement of the 30th ASEAN Ministers of Energy

Meeting (AMEM), 2014).[5] Other stated aims of the NEC-SSN include exchange of experts between countries and sharing of best technical and regulatory practices (Selvaraju, 2013).[6] However, the NEC-SSN remains in effect only a contact network among energy officials, and its activities thus fall short of what might be expected of a fully-functional nuclear safety regime.

In terms of setting up such a regime, one key difficulty is that ASEAN member states are most likely not fully convinced of the need to have an extensive nuclear safety regime at present, despite having committed to establishing one in principle. Although Indonesia, Malaysia, Thailand and Vietnam have all expressed interest in building nuclear power plants, they have postponed their plans by at least some degree following the Fukushima Daiichi accident. At this point, it will likely be at least a decade before ASEAN sees its first operational nuclear plant, most likely in Vietnam. ASEAN officials are thus far not convinced of the need to have an extensive regime on nuclear energy safety, or even to expand cooperation activities under the already-established NEC-SSN; there is not yet any strong sense of urgency.

In addition, ASEAN member states are already cooperating on a bilateral basis with countries that have mature nuclear energy programmes, such as Japan, South Korea, the United States and Russia, or have sought regulatory and technical advice directly from the IAEA. The IAEA manages an Asian Nuclear Safety Network (ANSN) that focuses heavily on practical concerns for countries seeking to build nuclear power plants. It conducts capacity-building workshops on issues such as emergency response,

---

[5] Association of Southeast Asian Nations (2012), *Joint Ministerial Statement of the 30th ASEAN Ministers of Energy Meeting (AMEM)*, Jakarta: ASEAN Secretariat, (http://www.asean.org/news/asean-statement-communiques/item/joint-ministerial-statement-of-the-30thasean-ministers-of-energy-meeting-amem; accessed 14 March 2014).

[6] Selvaraju, Mala (28 February 2013). ASEAN's Efforts Towards Non-Proliferation. In *20th Asian Export Control Seminar*. Tokyo, Japan: Ministry of Economy, Trade and Industry (http://www.meti.go.jp/english/press/2013/0228_04.html; accessed 14 March 2014).

operation of nuclear power plants, and the establishment of national regulations to address radiation risks ('Overview and Current Status of the Asian Nuclear Safety Network (ANSN), 2012.)'.[7] The existence of such bilateral and international cooperation lessens the immediate need for regional cooperation at the ASEAN level.

Is there a Need for an ASEAN Nuclear Safety Regime?

Many ASEAN countries are wary of committing to new international obligations. ASEAN favours a soft approach towards multilateral efforts rather than firm, legally binding agreements. The grouping also has a strong respect for national sensitivities and sovereignty, and nuclear safety is by definition a sensitive issue, as it ties directly into the national energy policy and long-term energy security of the involved countries.

National pride and prestige is also a potential obstacle for regional cooperation. Critics have suggested that some ASEAN countries are aiming to construct nuclear power plants as a competitive rush, seeking to gain status over their neighbours. As can be expected, the governments involved have so far denied that this is the case. Vietnam has characterised their country's development of nuclear power plants as increasing national standards of science and technology, but this is presented as an additional beneficial outcome rather than the leading reason to push for nuclear energy. That said, even if national pride is not the chief reason states are pursuing nuclear technology, it may be hard for countries to accept that their neighbours should have any insight and influence over their domestic energy plans.

However, the potential transboundary impact of nuclear energy production and potential accidents does elevate nuclear safety to a

---

[7] ANSN Management Team, Safety and Security Coordination Section, Department of Nuclear Safety and Security, IAEA (2012). Overview and Current Status of the Asian Nuclear Safety Network (ANSN). Asian Nuclear Safety Network (ANSN) (https://ansn.iaea.org/Common/WhatisANSN/documents/OverViewofANSN.pdf; accessed 14 March 2014).

regional issue. Only four of the ASEAN members have concrete plans to construct nuclear power plants; at the present time, the other six member states do not. However, even these remaining six countries share a vested interest in ensuring that plants constructed in this region are safe and secure. Given Southeast Asia's geography and the relatively small size of most ASEAN countries, a nuclear accident at a power plant in ASEAN would definitely result in environmental damage across national boundaries (Tay and Phir, 1998).[8] It is therefore logical that ASEAN establish mechanisms or plans for both crisis prevention and emergency response — setting up strong safety standards to prevent an accident from happening, and creating plans to deal with any accidents if they do arise.

Multilateral cooperation on nuclear safety would also help preserve and grow regional trust, an objective that is aligned with ASEAN's traditional role. Members of the public in ASEAN countries could potentially be concerned about nuclear plants being constructed in neighbouring states. If countries are seen to be working together under a regional nuclear safety regime, this would also help to reduce fears of a potential accident across the border.

In addition, ASEAN states seeking nuclear power face common and potentially costly challenges, such as the need for human resources training, establishment of national legislation to govern the operation of reactors, and eventual disposal of radioactive waste. Multilateral cooperation at the ASEAN level could be cost-effective for member countries, allowing states to pool resources to achieve greater economies of scale. Cooperation is feasible in three main areas, firstly in understanding regulatory issues, secondly in science and technology, and thirdly in human resources. ASEAN countries could give each other mutual assistance in implementing new IAEA safety standards, exchange engineering expertise or share best practices between national agencies, and create joint reactor operator training and certification programmes.

---

[8] Tay and Phir (1998), op. cit.

Regional ASEAN collaboration could also allow ASEAN to more efficiently interact with other countries, international organisations or companies, putting ASEAN members in a stronger position when dealing with third parties. For instance, one function of the European Atomic Energy Community (EURATOM) is to give EURATOM members greater control over the IAEA inspection process. Inspections continue to be conducted by the IAEA, but these are jointly organised with EURATOM, and European inspectors accompany the IAEA representatives. This reduces the administrative burden on the IAEA, while giving European nations a greater sense of confidence and ownership over the inspection process. ASEAN could likewise benefit from similar collective arrangements.

ASEAN has already begun to tentatively explore engagement as a collective grouping with the IAEA, for instance via the NEC-SSN, the ASEAN Regional Forum Inter-sessional Meeting on Non-Proliferation and Disarmament, and the ASEAN Secretariat. Notably, Mr. Yukiya Amano, Director-General of the IAEA, visited the ASEAN Secretariat in 2011 and met with the Committee of Permanent Representatives to ASEAN (CPR). These exchanges have thus far been limited to only meetings and briefings, but ASEAN is officially open to exploring more practical cooperation with the international agency (H.E. Le Luong Minh, 2013).[9] ASEAN officials do not see the IAEA as supplanting the need for ASEAN regional cooperation on nuclear safety. Likewise, cooperation at the ASEAN level is not viewed as a replacement for collaboration under the IAEA. Rather, the two could potentially complement each other.

---

[9]H.E. Le Luong Minh (12 February 2013). Speech by H.E. Le Luong Minh, Secretary-General of ASEAN. In Regional Seminar on Maintaining a Southeast Asia Free of Nuclear Weapons. Jakarta, Indonesia: ASEAN, ⟨http://www.asean.org/news/item/speech-by-he-le-luong-minh-secretary-general-of-asean-at-the-regional-seminar-maintaining-a-southeast-asia-region-free-of-nuclear-weapons-2; accessed 14 March 2014).

The Way Forward

ASEAN's NEC-SSN is only a modest step towards concrete regional cooperation on nuclear safety, but it is also a practical move for ASEAN. It would be out of character for ASEAN nations to rush into creating new international institutions. That said, having made successful first ventures in this area, it is important that ASEAN build on this momentum.

Several potential models have been suggested for the creation of a more formal regional nuclear safety regime in Southeast Asia. For instance, a wider "ASIATOM" modelled after EURATOM has been suggested by think tanks and policy analysts, including not only countries new to nuclear energy but Asian states with mature nuclear industries, such as China, Japan and South Korea. But a wider Asian organisation would face issues of membership; in the present political climate, a regional nuclear organisation bringing together China and Japan could be difficult to negotiate. Given these issues of membership, the creation of an ASEAN nuclear safety regime would be far more feasible in comparison, building on the grouping's existing efforts towards integration. ASEAN is a natural centre and hub for regional cooperation.

Regional cooperation on energy issues meshes with the current ASEAN agenda of building closer integration and connectivity, in line with the grouping's goal of establishing an ASEAN Community by 2015, with the three pillars of political security, economic, and socio-cultural cooperation. In 2007, ASEAN energy ministers vowed to link each country's power distribution networks into a regional grid. Under the ASEAN Power Grid Plan, Singapore is supposed to be fully connected to Johor and Batam by 2014. ASEAN also has an emergency petroleum-sharing scheme in case of oil shortages and plans to construct a Trans-ASEAN Gas Pipeline (TAGP) Network across the region. In 2010, the ASEAN members established the ASEAN Plan of Action for Energy Cooperation (APAEC), which includes the setting of targets for the development of renewable energy. In this context, regional cooperation on nuclear safety would be a logical move for ASEAN.

As discussed above, ASEAN's regional cooperation on nuclear issues has so far been more symbolic than substantial. In the short to medium term, ASEAN could further develop the existing NEC-SSN platform, concentrating on capacity building that includes human resources development, education and training, as well as the formulation of emergency preparedness and response plans. This capacity building could be conducted in cooperation and in conjunction with the IAEA, in order to facilitate the adoption of internationally recognised best practices and standards in the region (Murakami, 2012).[10]

In the long term, it may be viable for ASEAN to establish a more formal regional nuclear safety regime offering organised economic, legal and regulatory support to its member nations, with cost-sharing of resources, a common base of information and expertise, and other such schemes. In this, ASEAN could take cues from the example of EURATOM, as well as other initiatives in various regions. For instance, the Gulf Cooperation Council has conducted joint feasibility and implementation studies for nuclear power, the Baltic States and Poland have a nuclear development initiative, while Argentina and Brazil have held bilateral discussions on jointly building and operating reactors (Borovas and Teplinsky, 2011).[11]

Regardless of the model adopted, regional nuclear safety cooperation should ideally complement rather than compete with the global IAEA nuclear safety regime. ASEAN must also ensure any of its own regional arrangements are efficient, transparent and credible in order to be taken seriously. At the present time,

---

[10] Murakami Tomoko, (ed.) (2012). Study on International Cooperation Concerning Nuclear Safety Management in East Asian Countries. *ERIA Research Project Report*, 2012 (28), Jakarta: Economic Research Institute for ASEAN and East Asia, (http://www.eria.org/publications/research_project_reports/FY2012-no.28.html; accessed 10 March 2016).

[11] Borovas, George and Teplinsky, Elina (2011). ASEAN: The Next Nuclear Powerhouse?' *Infrastructure Journal*, 14 February 2011 (https://www.pillsburylaw.com/siteFiles/Publications/ASEANTheNextNuclearPowerhouse.pdf; accessed 14 March 2014).

several ASEAN countries have been criticised for not demonstrating sufficient commitment to international norms regarding nuclear safety.

Of the ASEAN countries that are close to actually building nuclear power plants, Indonesia and Vietnam have ratified all of the three major UN conventions on nuclear safety, with Vietnam's accession to two of the conventions coming only recently in 2012 and 2013. However, Thailand and Malaysia are still not party to all three conventions. Malaysia has signed but not ratified the UN Convention on Early Notification of a Nuclear Accident, and is not part of the UN Convention on the Physical Protection of Nuclear Material or the UN Joint Convention on the Safety of Spent Fuel Management and Safety of Radioactive Waste Management. Thailand is also not bound by the latter convention.[12] Participation in UN conventions is certainly not the only measurement of a state's commitment to nuclear safety, but they do serve as a strong signal of intentions that helps to reassure other nations. As ASEAN countries move closer to seeing their first nuclear reactors come online, it is important that they continue to send the right signals to the broader international community (Tay, Simon, S.C. *et al.*, 2011).[13]

Conclusion

Cooperation on nuclear safety is a natural fit with ASEAN's progress towards becoming a regional community. At the same time, it is worth noting that proposals to create a regional nuclear safety regime, whether more broadly across the Asia-Pacific or only within ASEAN, have been discussed by national representatives and academics for nearly two decades, but actual progress has been slow. The challenges involved in creating a regional nuclear safety regime cannot be underestimated. However, as several ASEAN countries, starting with Vietnam, may have operational

---

[12]Status of convention parties retrieved from IAEA website.
[13]Tay, Simon S.C. *et al.* (2011). Examining Nuclear Energy Ambitions in Southeast Asia. *SIIA Special Report*, 11(02). Singapore: SIIA.

nuclear power plants within the next two decades, there is an urgent need to promote proper practices in Southeast Asia. If ASEAN countries are indeed intending to build nuclear power plants, it is in the region's collective interest to ensure that these plants are constructed and operated to the highest standard, in a transparent fashion that is able to stand up to both public scrutiny from Asian citizens, as well as international attention from across the world (Tay, Simon, S.C. *et al.*, pp. 41–44, 2011).[14]

States cannot afford to be passive in this respect, and must respond pro-actively to ensure nuclear energy plans proceed with due care and caution. Although Singapore and many other nations in ASEAN have no current plans to run their own nuclear reactors, they do have a vested interest in ensuring that plants constructed in this region are safe and secure.

In addition, with ASEAN states moving towards connecting their power grids and transferring electricity across borders, all countries stand to benefit if successfully-run nuclear plants are created and operated.

While Singapore and other countries may not plan on hosting their own nuclear reactors, there are ways they can be involved, such as helping with technical expertise and capacity building. At the same time, it is important for them to have experts who are able to monitor and assess the regional nuclear developments as they progress in the years ahead. Right now, there is no widespread sense of urgency over nuclear safety and other related issues in ASEAN. For the time being, it seems the prospect of operational nuclear power plants in ASEAN is sufficiently far-off for countries to take a "wait and see" attitude. But it is important that both governments and the public start discussing key issues such as safety standards, and how to respond collectively in the event of an actual emergency. The March 2011 accident at Fukushima Daiichi underscores an important lesson on the dangers of complacency that ASEAN cannot afford to ignore.

---

[14] Ibid.

Bibliography

ASEAN (1995). Treaty on the Southeast Asia Nuclear-Weapon-Free-Zone, 15 December 1995. Jakarta: ASEANWeb; accessed 14 March 2014.

IAEA. Convention on Early Notification of a Nuclear Accident (Convention Status). *International Atomic Energy Agency (IAEA)*. IAEA, 17 September 2013. (Web; accessed 17 March 2014).

IAEA. Convention on the Physical Protection of Nuclear Material (Convention Status). *IAEA*. IAEA, 17 December 2013. (Web; accessed 17 March 2014).

IAEA. Joint Convention on the Safety of Spent Fuel Management and on the Safety of Radioactive Waste Management (Convention Status). *IAEA*. IAEA, 9 October 2013. (Web; accessed 17 March 2014).

Chapter 3

AREAS OF REGIONAL NUCLEAR ENERGY COOPERATION

Eulalia Han[1]

Introduction

The March 2011 Fukushima nuclear accident reinvigorated discussions over the safety and security of nuclear energy. It influenced governments to relook at their current nuclear energy aspirations, nuclear program, safety standards and emergency preparedness framework. Issues of regional nuclear energy cooperation are also brought to the fore as states re-examine and re-emphasise the significance of collective action in managing a nuclear energy disaster as the latter's consequence extend beyond national borders. Since Fukushima, there have been major shifts in the attitudes of many European countries towards their respective countries' nuclear energy program. This is reflected in Germany's, Italy's and Switzerland's decision to decommission all reactors. However, nuclear energy continues to be an attractive option to many countries that remain concern over unstable oil prices, the reliability of renewable energy and carbon emissions.

---

[1] Research Fellow, Energy Studies Institute, National University of Singapore.

This chapter provides a brief overview of Southeast Asia's nuclear energy ambitions before looking at the potential areas for regional nuclear energy cooperation. While much has been said on the various aspects of nuclear energy that Southeast Asia can cooperate on, these suggestions do not usually consider the regional framework, existing energy infrastructure and current realities that Southeast Asian states are a part of. The Association of Southeast Asian Nations (ASEAN), the regional organisation that Brunei, Cambodia, Indonesia, Laos, Malaysia, Myanmar, the Philippines, Singapore, Thailand and Vietnam are members of, operates on the norms of non-interference in the domestic affairs of member states, peaceful settlement of all disputes, regional autonomy, and the practice of the "ASEAN Way" (Acharya, 2005, pp. 98–99).[2] ASEAN's preference for informality and consensus means that "security has always been addressed through consultation and dialogue rather than through conventional collective security and formal mechanisms for settling disputes" (Leifer, 1996, p. 14).[3] Therefore, it is important to understand the extent of regional nuclear energy cooperation within the aforementioned principles so as to provide a realistic and fair analysis of prospective areas of cooperation.

Southeast Asia's Nuclear Energy Ambitions

Southeast Asia's interest in nuclear power stems from its energy security concerns. As the region continues to experience rapid economic growth, the need to diversify its sources of energy, while reducing its dependence on fossil fuel imports, has steered the region to consider the possibility of integrating nuclear energy into the region's energy mix. In addition, "prestige", "economic competition", "regional influence" and the continued production of radioisotopes for use in the medical and agricultural sectors

---

[2] Acharya, Amitav (2005). Do Norms and Identity Matter? Community and Power in Southeast Asia's Regional Order. *The Pacific Review* 18(1), 95–118.
[3] Leifer, Michael (1996). The ASEAN Regional Forum, *Adelphi Paper No. 302*. London: International Institute for Strategic Studies.

further motivates ASEAN Member States' nuclear ambitions as they follow closely the nuclear energy plans of their neighbours (CNS, CENESS and VCDNP, 2012, p. 3).[4]

Vietnam leads the region in the development of nuclear energy as it remains the only country with an active nuclear program. The Vietnamese government has signed an agreement with Rosatom, Russia's state nuclear energy company, for Russia's assistance in the construction of the Ninh Thuan 1 plant that will consist of two Russian-designed pressurised water reactors with a total nuclear capacity of 2000 megawatt electric (MWe) (World Nuclear Association, 2014).[5] Construction of the Ninh Thuan 1 plant was initially scheduled to begin in 2014 but as government officials had raised needs to further enhance safety and human resource training, there would be a minimum of a two years delay (Vietnamnet, 2014).[6] Vietnam has also signed an intergovernmental agreement with Japan to design, construct and operate a second nuclear power plant. The design of the technology is still pending since the results of the feasibility study that concluded in May 2013 (World Nuclear Association, 2014).[7] Vietnam is also in the process of a joint feasibility study with South Korea regarding the possibility of engaging Korean nuclear technology in the future.

According to the latest revised data from the World Nuclear Association on 3 January 2014, Vietnam has planned for four

---

[4]CNS (James Martin Center for Nonproliferation Studies), CENESS (Center for Energy and Security Studies) and VCDNP (Vienna Center for Disarmament and Non-Proliferation) (2012). Nuclear Development in Southeast Asia. In *Prospects for Nuclear Security Partnership in Southeast Asia*. Monterey/Moscow/Vienna: CNS, CENES and VCDNP.

[5]World Nuclear Association (2014). Nuclear Power in Vietnam. *World Nuclear Association*, (http://www.world-nuclear.org/info/Country-Profiles/Countries-T-Z/Vietnam/; accessed 26 January 2014).

[6]Vietnam Delays Construction of the First Nuclear Power Plant (20 January 2014). *Vietnamnet* 2014, (http://english.vietnamnet.vn/fms/science-it/94169/vietnam-delays-construction-of-the-first-nuclear-power-plant.html; accessed 26 January 2014).

[7]World Nuclear Association, op.cit.

reactors with a total capacity of 4000 MWe and six proposed reactors, totalling 6700 MWe. Indonesia has also planned for two reactors (2000 MWe) with four proposed reactors, totalling 4000 MWe in nuclear capacity. While Thailand has no current plans for any reactors, the country has also proposed five reactors (5000 MWe) in the future. In addition, Indonesia, the Philippines, Thailand and Vietnam have also used Highly Enriched Uranium (HEU), uranium with more than 20 per cent of U-235 isotopes, for their research reactors (CNS, CENESS and VCDNP, 2012, p. 15).[8]

Current Regional Nuclear Energy Cooperation

There are current regional initiatives in which ASEAN Member States are a part of. As an extension of the declaration for a Zone of Peace, Freedom and Neutrality (ZOPFAN) amongst ASEAN states in 1971, the Southeast Asia Nuclear-Weapon-Free Zone (SEANWFZ) treaty was established in 1995 and ratified in March 1997. As reiterated by ASEAN's Secretary General H.E. Le Luong Minh,

> "By signing the SEANWFZ Treat in December 1995, the ASEAN Member States committed themselves to nuclear disarmament and non-proliferation in the region. This commitment was further reaffirmed with the entry into force of the ASEAN Charter which has, as one of its goals, the preservation of Southeast Asia as a zone free of nuclear weapons and free of all other weapons of mass destruction." (H.E. Le Luong Minh, 2013).[9]

Member States also participate in various forums and initiatives such as the Asian Nuclear Safety Network (ANSN), the

---

[8] CNS (James Martin Center for Nonproliferation Studies), CENESS (Center for Energy and Security Studies) and VCDNP (Vienna Center for Disarmament and Non-Proliferation), op.cit.

[9] H.E. Le Luong Minh, (12 February 2013). Speech by H.E. Le Luong Minh Secretary General of ASEAN at The Regional Seminar: *Maintaining a Southeast Asia Region Free of Nuclear Weapons*, (http://www.asean.org/news/item/speech-by-he-le-luong-minh-secretary-general-of-asean-at-the-regional-seminar-maintaining-a-southeast-asia-region-free-of-nuclear-weapons-2; accessed 24 January 2014).

Forum for Nuclear Cooperation in Asia and the Asia-Pacific Safeguards Network. ASEAN Energy Ministers have also established the Nuclear Energy Cooperation Sub-Sector Network (NEC-SSN) in 2010:

> "to forge a region-wide cooperation on the use of nuclear energy for power generation purposes through a broad exchange of informa-tion and assistance on safe and sustainable civilian nuclear power programmes" (Pitsuwan, 2011).[10]

Vietnam is the current network coordinator for the NEC-SSN. Brunei, Cambodia, the Philippines, Singapore and Thailand have also participated in capacity building activities that are related to the Proliferation Security Initiative (PSI), a multinational response to the challenge posed by the proliferation of weapons of mass destruction.

Another Memorandum of Understanding (MoU) that could potentially have an acute impact on future regional nuclear energy cooperation is the MoU on the ASEAN Power Grid (APG) that looks to "achieve long-term security, availability and reliability of energy supply" and harmonise "all aspect of technical standard and operating procedure as well as regulatory frame works among member country" (ASEAN Centre for Energy, (2010)).[11] The even-tual goal of this MoU is to totally integrate Southeast Asia's power grid system. The potential challenge could possibly be the unease that nuclear-powered countries feel when non-nuclear powered countries in the region are connected to their electricity grid, thus,

---

[10]H.E. Dr Surin Pitsuwan (May 2011), Keynote Speech by H.E. Dr. Surin Pitsuwan, Secretary-General of ASEAN, at the *2011 Asia Pacific Public Policy Forum on Energy, Innovation and Sustainable Development*, Jakarta, (http://www.asean.org/resources/2012-02-10-08-47-56/speeches-statements-of-the-former-secretaries-general-of-asean/item/keynote-speech-by-dr-surin-pitsuwan-secretary-general-of-asean-at-the-2011-asia-public-policy-forum-on-energy-innovation-and-sustainable-development; accessed 26 January 2014).

[11]ASEAN Centre for Energy (2010). Program Areas: Program Area No.1 ASEAN Power Grid. *ASEAN Plan of Action for Energy Cooperation 2010–2015*. Jakarta: ASEAN Centre for Energy.

benefitting or having a stake in the former's nuclear program. The next section will discuss these issues further.

Challenges to Regional Nuclear Energy Cooperation

Michael Malley, a specialist in Southeast Asian politics at the Naval Postgraduate School in the United States, observes that Southeast Asian nations are reluctant to cooperate on nuclear energy and see greater benefit in collaborating with international institutions such as the International Atomic Energy Agency (IAEA) than ASEAN regarding issues related to nuclear energy (Malley, 2008, pp. 241–56).[12] Although existing frameworks exists in ASEAN that are relevant to ensuring the safety and security of nuclear energy, Member States have appeared unwilling to engage these initiatives. Therefore, beneath

"the rhetoric of consensus, there is apparently little agreement among ASEAN members that cooperation within Southeast Asia on nuclear safety and security is a desirable adjunct to international cooperation through the IAEA on these matters" (IISS, 2009, p. 12).[13]

Malley (2008) also observes that only Singapore and the Philippines are encouraging the development of ASEAN institutions to deal with nuclear energy issues, but that the more determined countries that are seriously considering nuclear energy prefer to work with the IAEA instead.[14] Therefore, the clash of short-term

---

[12] Malley, Michael S. (2008). Bypassing Regionalism? Domestic Politics and Nuclear Energy Security. In *Hard Choices: Security, Democracy, and Regionalism in Southeast Asia*. D. K. Emmerson (ed.) pp. 241–262, Stanford: The Walter H. Shorenstein Asia-Pacific Research Center.

[13] International Institute for Strategic Studies (IISS) (2009). Chapter 1: Regional Cooperation. In *Preventing Nuclear Dangers in Southeast Asia and Australasia*. IISS Strategic Dossier, pp. 11–19. London: International Institute for Strategic Studies.

[14] Malley, op.cit, pp. 241–56.

and long-term plans of the respective Southeast Asian nations poses a major obstacle in regional nuclear energy cooperation.

Relating back to potential challenges that could result from the total integration of Southeast Asia's power grids on future nuclear energy cooperation, Singapore's expected connection to the national grids of Malaysia (Johor) and Indonesia (Batam) by 2014, for example, would make it possible for the former to receive electricity generated by nuclear power from plants that are within proximity from the city state (IISS, p. 14).[15] Malley (2008) argues that:

> "Singapore, a party with no nuclear plans of its own but substantial concerns about the safety and security of prospective nuclear plants in the region... had discussed the possibility of gaining "leverage" over its neighbo[u]rs' nuclear safety policies by integrating its electricity grid with theirs".[16]

Therefore, both Indonesia and Malaysia may view with unease Singapore's connection to its grid should they decide to include nuclear power in its energy mix as the latter could effectively influence and have a stake in Kuala Lumpur's and Jakarta's civilian nuclear program (IISS, 2009, pp. 14–15).[17]

Future nuclear energy cooperation could also be hindered by existing territorial disputes such as claims over the South China Sea (Raine and Le Miere, 2013, pp. 105–150),[18] border clashes between Thailand and Cambodia (Sothirak, 2013, pp. 87–90),[19] and the sultanate of the southern Philippines province of Sulu's

---

[15] IISS, op.cit, p. 14.

[16] Malley, op.cit, p. 255.

[17] IISS, op.cit, pp. 14–15.

[18] Raine, Sarah and Le Miere, Christian (2013). *Regional Disorder: The South China Sea Disputes*. New York: The International Institute for Strategic Studies (IISS).

[19] Sothirak, Pou (2013). Cambodia's Border Conflict with Thailand. In *Southeast Asian Affairs 2013*, Daljit Singh (ed.), pp. 87–100, Singapore: Institute of Southeast Asian Studies.

historical claim over Malaysian state of Sabah (Whaley, 2013).[20] In addition, Tanya Ogilvie-White (2006) also concludes that with "the exception of Singapore, Southeast Asia's export control systems remain unsophisticated and weak" (p. 10).[21] Maintaining a well-regulated system is important in preventing the sale or transfer of sensitive technology to rogue regimes (Parameswaran, 2009, p. 5).[22]

Potential Areas of Nuclear Energy Cooperation

Policymakers and scholars have suggested many areas in which Southeast Asia can cooperate on. These areas include exchanging information on technological know-how, establishing joint regional training centres (CNS, CENESS and VCDNP, 2012, p. 46),[23] the development of an organisation within the ASEAN umbrella that is equivalent to the European Atomic Energy Community (EURATOM) (Symon, 2008, pp. 130–33),[24] ratifying major international nuclear security conventions (Boureston and Ogilvie-White, 2010, pp. 2–5),[25] and encouraging regional cooperation in emergency mitigation,

---

[20] Whaley, Floyd, Sabah Claim Hits Malaysia, Philippines (4 March 2013), *The Sydney Morning Herald*, (http://www.smh.com.au/world/sabah-claim-hits-malaysia-philippines-20130304-2ffmz.html?skin=text-only; accessed 30 January 2014).

[21] Ogilvie-White, Tanya (2006). Non-proliferation and Counter-terrorism Cooperation in Southeast Asia. *Contemporary Southeast Asia*, 28(1), 1–26.

[22] Parameswaran, Prashanth (2009). Southeast Asia's Nuclear Energy Future: Promises and Perils. Project 2049 Institute, (https://project2049.net/documents/southeast_asia_nuclear_energy_future.pdf; accessed 30 January 2014).

[23] CNS, CENESS and VCDNP, op.cit, p. 23.

[24] Symon, Andrew (2008). Southeast Asia's Nuclear Power Thrust: Putting ASEAN's Effectiveness to the Test? *Contemporary Southeast Asia* 30(1), pp. 118–39.

[25] Boureston, Jack and Ogilvie-White, Tanya (2010). Seeking Nuclear Security through Greater International Coordination, Working Paper. *International Institutions and Global Governance Program*, New York: Council on Foreign Relations.

preparedness and response (Murakami, 67–93).[26] These are valid suggestions and would be necessary as part of the region's aim to establish a well-developed nuclear energy program. However, even before these suggestions can mature, the focus of Southeast Asia's future cooperation on nuclear energy should first be about agreeing on a common conceptual understanding of nuclear security, advancing a more coherent and comprehensive safety culture, and understanding the extent to which each country would want to cooperate given the context of existing energy infrastructures such as the APG while recognising that not all countries would have an active nuclear program.

Having greater conceptual clarity of what nuclear security entails would require ASEAN Member States to be a signatory to every regional or multilateral nuclear security convention that the other has ratified. This would allow

"a clearer understanding of what is 'in' or 'out' of the core nuclear security framework, especially in relation to plurilateral initiatives with questionable mandates, such as GICNT [Global Initiative to Combat Nuclear Terrorism], and how they relate to the IAEA nuclear security program" (Kidd, Cottee and Miller, 2012, p. 33).[27]

It also allows for the streamlining of regional efforts. If Southeast Asian nations are not able to reach a consensus on the various international conventions to ratify, there could be a loophole to be exploited, resulting in the constant negotiation on the

---

[26] Murakami, Tomoko (ed.) (2013). Study on International Cooperation Concerning Nuclear Safety Management in East Asian Countries, *ERIA Research Project Report 2012, No. 28*, (http://www.eria.org/RPR_FY2012_No.28_front_cover.pdf; accessed 30 January 2014).

[27] Kidd, Joanna, Cottee, Matthew and Milller, Carl (2012). Nuclear Security in Southeast Asia. Prepared for the Carnegie Corporation of New York. *International Centre for Security Analysis (ICSA) King's College London (KCL)*, (http://www.kcl.ac.uk/sspp/policy-institute/icsa/Southeast-Asia-Nuclear-Security---KCL-(ICSA)-Jan-12.pdf; accessed 30 January 2014).

terms to be agreed upon. Countries without an active nuclear program should also be signatories to the same conventions so as to ensure the harmony of legal frameworks and emergency plans.

An extension of the harmonisation of each country's legal and emergency management framework would be to emphasise the adoption of a coherent and comprehensive safety culture that is not only shared between states, but cultivated in society (Chew, 2010).[28] Strengthening disaster plans and building resilience amongst citizens is important. Integrating aspects of nuclear hazards into current emergency preparedness programs is also necessary in preparing the citizenry for a future in which nuclear energy is a part of the region's energy mix (Jamil and Gong, 2013).[29] Therefore, public education has been identified as key to developing a comprehensive safety culture. However, Southeast Asian societies, though it might not be specific to this region, have not been well-informed on both the challenges and opportunities of nuclear energy, much less the procedures to follow in times of an emergency. As the region plans to include nuclear power in its energy mix, public education would be fundamental as part of a well-considered nuclear program.

The last issue that warrants extensive discussion is understanding the extent to which ASEAN Member States see nuclear energy cooperation as necessary and beneficial to all countries, especially when considering other ongoing plans such as the APG that would connect the region's power grids. As noted, not all ASEAN Member States are equally enthused over regional cooperation as the

---

[28] Chew, Alvin (2010). Nuclear Energy in Southeast Asia: Competition or Cooperation? RSIS Commentaries, S. Rajaratnam School of International Studies, (RSIS), Nanyang Technological University, (http://www.rsis.edu.sg/rsis-publication/nts/1456-nuclear-energy-in-southeast-as/#.VmNAcSvoZjM; accessed 30 January 2014).

[29] Jamil, Sofiah and Gong, Lina (2013). Nuclear Energy Development in Southeast Asia: Implications for Singapore. NTS Insight, S. Rajaratnam School of International Studies (RSIS), Nanyang Technological University, (https://www.rsis.edu.sg/rsis-publication/nts/2652-nuclear-energy-development-in/#.VmNBhCvoZjM; accessed 30 January 2014).

countries with more concrete plans to develop a nuclear program prefer instead to consult directly with the IAEA. Moreover, the APG would allow all countries to receive electricity generated from nuclear power, including countries with no operating reactors. Nuclear-powered states might then perceive themselves to be at a disadvantaged position as the region gets to benefit while they bear the brunt of direct costs and risks associated with nuclear power. Therefore, it is essential to develop "an ASEAN approach to nuclear power development [that could] mesh with other aspects of cooperative ASEAN energy programmes, in particular efforts to foster cross-border electricity transmission supply" (Symon, 2008, p. 133).[30] Regional nuclear energy cooperation, therefore, should be discussed within existing energy infrastructures. Only then would the opportunities and limitations of cooperation surface, especially the extent to which the respective ASEAN states value regional nuclear energy cooperation, and in the areas of cooperation that would not violate the region's practice of non-interference and sovereign rights.

Conclusion

There have been suggestions in the past that contend that a common ASEAN nuclear power plan could prove more advantageous than a situation where we have ten states with ten individual nuclear programs. An ASEAN-operated nuclear power plant could see Member States contribute to

> "the common goal of attaining reliable energy", reduce greenhouse emissions and "reduce the duplication of technology, but also foster individual nations to develop expertise in their niche areas... [that] could lead to more advanced technology being adopted at a cheaper cost" (Chew, 2010).[31]

---

[30] Symon, op.cit, p. 133.
[31] Chew, op.cit.

However, with Vietnam having signed bilateral agreements with both Russia and Japan, together with Indonesia, Thailand and the Philippines who seem interested in developing their own nuclear programmes, a common ASEAN-operated nuclear power plan would, at most, be a long-term goal.

ASEAN can follow EURATOM's lead on past and present areas of nuclear energy cooperation. Existing research and discussions on Southeast Asia's cooperation on nuclear-related matters have tended to steer towards similar arguments. While this is useful as a sort of guidance to accompany the region's nuclear program, they can only bear fruit once Southeast Asia agrees to some fundamentals such as establishing a common understanding of nuclear security, adopting a safety culture that involves public education on nuclear energy and discussing areas of cooperation within the context of existing energy infrastructures. Only then can Southeast Asian nations build a strong foundation for their respective states' future nuclear energy program and move on to discuss the areas of regional nuclear energy cooperation that are presently being considered such as establishing joint training centres, exchange of technological know-how, and the sharing of best practices.

Chapter 4

CHALLENGES IN ADAPTING MALAYSIA'S LEGAL FRAMEWORK TO INTERNATIONAL NUCLEAR NORMS

Aishah Bidin[1]

Introduction

Similar to many countries worldwide, Malaysia, like many fast developing Asian countries, is considering the introduction of nuclear power as a part of the energy mix. This would allow for both the diversification of key energy supplies and also assure the availability of low cost energy that is essential for sustained growth. Moreover, nuclear energy is the only large-scale source of energy that is not expected to be penalised by the (inevitable) pricing of carbon, which is expected to be introduced in the future. In addition, Malaysia's rapid economic progress over the last decades has brought the country to the level of technological development and financial strengths where adding nuclear energy into the mix would become a highly interesting consideration.

---

[1] Dean of the Law Faculty, National University of Malaysia.

Malaysian Legal and Regulatory Nuclear Framework

Reflecting on this, in June 2009, the Government of Malaysia made a decision to actively pursue the consideration of nuclear power as an alternative source of electric energy, for the period post 2020. In this, the Government established an organisation called the Malaysia Nuclear Power Corporation ("MNPC"), intended on being at the forefront of the activities. The current development plans call for the commissioning of the first NPP in about 11 to 12 years. While this is an ambitious plan, for a country with the developed technical and financial infrastructure that Malaysia has, with a dedicated population and outstanding professionals behind the activity, this schedule is definitely achievable.

Malaysia does not have direct activities supporting nuclear energy, but some of the core experiences with nuclear power exist in the country. This includes the operation of a research reactor, other research and development activities, as well as activities within the Atomic Energy Licensing Board (AELB) and wider, within the Ministry of Science, Technology and Innovation (MOSTI). Malaysia regularly participates in a variety of activities supported by the International Atomic Energy Agency (IAEA). While having nuclear experience and competence is an important element when deciding to embark on the nuclear energy path, an essential pre-condition is the establishment of the appropriate infrastructure that will assure that safety is seen as the utmost priority. In the post- Fukushima nuclear world, this would become even more strongly pronounced, as it can be seen from the announcement issued by an important nuclear country, saying that "the exports of nuclear technology will be undertaken only to countries with adequately developed nuclear infrastructure".

The introduction of nuclear energy in any country, Malaysia included, requires a long lead time, with concerted and comprehensive planning and development of appropriate institutional, legal and regulatory, technological, educational and training, and other related infrastructure. Among the key requirements are the need to establish the economic viability of nuclear power projects, a long-term national commitment for the nuclear energy option, a national pre-selection of the appropriate nuclear power plant type, a long-term spent-fuel management policy, compliance with the international system of nuclear governance, national capacity-building and public acceptance programme on nuclear energy.

In pursuant to the decision by the Government to consider nuclear energy as an option for electricity generation, especially in the Peninsula, post-2020, a national Nuclear Power Development Steering Committee (JKPPN), chaired by the Secretary General of the Ministry of Energy, Green Technology and Water (KeTTHA), has been established. Three Working Committees, comprising of relevant Ministries, Government agencies and Government-Linked Companies (GLCs), have also been established, as follows;

i. Nuclear Power Programme Development Working Committee, led by the Malaysian Nuclear Agency (Nuclear Malaysia);
ii. Nuclear Power Project Development Working Committee, led by Tenaga Nasional Berhad (TNB), as the electric utility for Peninsular Malaysia; and,
iii. Nuclear Power Regulatory Development Coordination Working Committee, jointly-led by the Atomic Energy Licensing Board (AELB) and Energy Commission (ST).

This structure comprising of the JPPKN and its three Working Committees can effectively be considered as a national Nuclear Energy Programme Implementation Organisation (NEPIO). As the lead agency for the Nuclear Power Programme Development Working Committee, one of the activities that Nuclear Malaysia is tasked with is Public Acceptance (PA) programme. As such, PA is essential in the area of nuclear power and needed to dispel myths regarding nuclear energy, including negative preconceptions, distorted and biased assertions, and exaggerated information about the risks of nuclear energy. PA also helps the public to understand that nuclear power generation is a useful energy resource which drives a nation's economic growth and builds a positive social harmony in the lives of the people.

To educate the public about nuclear science and technology and additionally, to respond to misinformation, there is a need for Malaysia to develop a comprehensive national information management database. This database will contain pertinent material ranging from basic to detailed information related to issues in nuclear science and technology. It will provide open and quick access to accurate, transparent and credible information for all stakeholders.

The support from IAEA is needed to provide assistance in devising and structuring the backbone of the NIDD: the information itself. Experts are required to guide the NIDD Management Team in filtering, sorting, classifying and categorising relevant information; authenticating the validity of information; safeguarding the database from malicious intent; and making its interface creative and interactive, while maintaining its user-friendliness and accessibility at the same time. The NIDD Management Team can consist of data collectors, writers, IT specialists, web designers, and cyber security personnel from different agencies.

International Instrument Relevant for Nuclear Power Programme

If Malaysia moves forward with a nuclear power programme, adherence to and implementation of a number of these instruments will be important in enabling Malaysia to obtain necessary cooperation from relevant suppliers. Also, participating in the review process under several of these instruments can ensure that Malaysia's programme meets recognised standards for safety, security, safeguards and liability.

These instruments fall into two categories. First, there are so-called legally binding or "hard law" instruments that establish concrete obligations and relationships under international law. Instruments such as conventions, treaties or agreements of a multilateral, regional or bilateral character fall within this category. They typically require negotiation, signature and approval or ratification under a state's constitutional arrangements. A second category of so-called "soft law" instruments are not considered legally binding or do not establish concrete obligations under international law. This category includes guidance or advisory documents developed by relevant international bodies, such as the IAEA.

Codes of Conduct are a common form of "soft law" instrument, as are various IAEA standards documents and technical documents. These instruments can be voluntarily applied by nations utilizing nuclear energy as a reflection of "best practice" in this highly technical field. Soft law instruments can help in harmonising and regularising the control activities of different states, even if they do not establish binding legal obligations. Implementation of these instruments can also demonstrate compliance with hard law instruments. However, "soft law" instruments can be converted into "hard law" in two ways. First, they can become a requirement for approval of IAEA project agreements under the Agency's technical cooperation rules. Second, they can be adopted as requirements or license conditions under domestic law.

With regard to so-called "hard law" instruments to which Malaysia has adhered, Malaysia obviously already has a duty to implement those instruments, including the adoption of national legislation, where applicable. For instruments that Malaysia has signed, but not yet completed domestic ratification procedures (such as the Additional Protocol for IAEA Safeguards), Malaysia has the sovereign right to take actions that would be consistent with the "objects and purposes" of such instruments. Finally, there are other instruments that Malaysia has not yet joined that deal with important aspects of nuclear technology. Some of these instruments have attained a broad level of acceptance by the international nuclear community, and can be said to represent best practice or useful guidance on which Malaysia may wish to utilise in framing its revised nuclear legislation. In addition, Malaysian adherence to the key instruments discussed in this paper would contribute to a growing international consensus on the necessary legal framework for the peaceful utilisation of nuclear energy and ionising radiation, thereby contributing to the development of nuclear law.

Specific "hard law" instruments reviewed in this paper include:

- The Convention on the Physical Protection of Nuclear Material (CPPNM) (1987)
- Amendment of the Convention on the Physical Protection of Nuclear Material and Nuclear Facilities (2005)
- Convention on Nuclear Safety (CNS) (1994)
- Joint Convention on the Safety of Spent Fuel Management and on the Safety of Waste Management (1997)
- Convention on Early Notification of a Nuclear Accident (1986)
- Convention on Assistance in the Case of a Nuclear Accident (1986)
- Treaty on the Non-Proliferation of Nuclear Weapons (NPT) (1968)
- Southeast Asia Nuclear Weapons Free Zone Treaty
- Safeguards Agreement with IAEA in connection with the NPT (INFCIRC/182)
- Model Protocol Additional to the Agreements between States and the IAEA for the Application of Safeguards (INFCIRC/540) (1997)
- International Convention for the Suppression of Acts of Nuclear Terrorism (CNT) (2005)
- Civil Liability Instruments
- UN Security Council Resolution 1540 (28 April 2004)

Specific "soft law" instruments presented in this paper include:

- Code of Conduct on the Safety and Security of Radioactive Sources (2003)
- Code of Conduct on the Safety of Research Reactors (2004)
- Physical Protection of Nuclear Material (INFCIRC.225/Rev. 5) (2011)
- Governmental, Legal and Regulatory Framework for Safety, IAEA General Safety Requirements Part 1 (2010)

Challenges in Adherence to International Legal Instrument

Adherence to relevant international instruments can have a positive impact on public opinion in Malaysia and by stakeholders in other States (particularly those in geographical proximity to Malaysia) that Malaysia is taking responsible actions to ensure that its programme meets accepted international standards.

The Convention on the Physical Protection of Nuclear Material from 1987 ("CPPNM"), its Amendment from 2005 ("CPPNM/A") and the International Convention for the Suppression of Acts of Nuclear Terrorism from 2005 ("ICSANT") stand out as critically important international conventions specific to nuclear security.[2]

Together with nuclear safety conventions, safeguards agreements established according to the Non-Proliferation Treaty and non-binding codes of conduct, the CPPNM, CPPNM/A and ICSANT have created a platform of international legal instruments upon which national law, regulations and institutional capability to govern safe, secure and peaceful use of the atom might be constructed.

In general, incorporation into national law and implementation of international counter-terrorism instruments pertaining to nuclear security do not occur automatically. States vary in how they accomplish incorporation and implementation and in their need to create new legislative, regulatory or institutional

---

[2]The International Atomic Energy Agency's Office of Legal Affairs ("IAEA/OLA") has prepared an *"Unofficial English Version of the Text of the [CPPNM], Adopted on 26 October 1979, Reflecting the Amendment Adopted by the States Parties to the Convention on 8 July 2005"* ("Unofficial English text of the amended CPPNM"). The text can be downloaded by using the following URL: http://ola.iaea.org/OLA/documents/ACPPNM/Unofficial%20consolidated%20text-English.pdf. The CPPNM and ICSANT have entered into force. The CPPNM/A has not yet entered into force. The IAEA/OLA document anticipates what the "Convention on the Physical Protection of Nuclear Material and Nuclear Facilities" will look like once the CPPNM/A enters into force.

infrastructure to adequately incorporate and implement new requirements. However, almost assuredly, states turn to their existing law as the starting point for deciding how they will give operative effect to the obligations they will be undertaking in adopting an international instrument and in carrying out those obligations. A number of factors influence these choices, including a state's system of law and its organisation, its existing legislative, regulatory and institutional frameworks for oversight, regulation and implementation, and the political process in which it must engage in order to complete adoption and implementation of the instrument.[3]

The scope of a state's existing programme and applications for the peaceful uses of nuclear energy would also influence the choices. In addition, the obligations that UNSCR 1540 imposes on all States pursuant to Chapter VII of the United Nations Charter require each State to have in place legal, regulatory and institutional infrastructure sufficient that it can meet such obligations as they might pertain to combating nuclear terrorism.[4]

a) *Approval*

Obtaining approval from relevant national authorities, including the legislature, can be a time-consuming process that involves a substantial commitment by Malaysian officials. Participating in the review meetings conducted pursuant to several instruments can involve a substantial effort by Malaysian authorities in preparing

---

[3]UNODC's publication, "Legislative Guide to the Universal Legal Regime Against Terrorism," provides an excellent summary of steps to complete adoption and implementation (pp. 10–11). Copies of the entire publication can be downloaded from the UNODC website: http://www.unodc.org/documents/terrorism/Publications/Legislative_Guide_Universal_Legal_Regime/English.pdf.

[4]The IAEA's Handbook on Nuclear Law is particularly useful to states who need to establish such infrastructure from the start. To view or download a copy of the first volume, see: http://www-pub.iaea.org/MTCD/publications/PDF/Pub1160_web.pdf. For implementing legislation, see volume 2 of the Handbook: http://www-pub.iaea.org/MTCD/publications/PDF/Pub1456_web.pdf.

a national report, reviewing the reports of other parties and participating in review meetings. Any necessary formal approval of the Government would follow. Then a formal communication from the Government of Malaysia would need to be submitted to the relevant depositary for that convention or treaty.

For the binding instruments developed under the aegis of the IAEA (such as the CPPNM, CNS, Joint Convention or Vienna Convention), a communication would be needed to the Director General from the relevant Malaysian authority, presumably through a diplomatic note transmitted by the Permanent Mission of Malaysia to the IAEA or International Organizations in Vienna. For instruments developed under the United Nations (such as the CNT) the communication would need to be submitted to the Secretary General through Malaysia's Permanent Mission in New York.

Non-binding guidance documents do not typically involve any formal indication by a state that it is applying them in national law. Rather, a statement of policy may be submitted in whatever forum may be appropriate, such as technical meetings of the IAEA. However, the Code of Conduct on Radioactive Sources does have a somewhat more structured procedure under which the IAEA Board has requested that states inform the IAEA Director General of their decision to apply the code in domestic activities utilising radioactive sources.

b) *Timing*

A question that arises in determining how Malaysia should approach the process of adhering to the various instruments is the schedule for such action. It is recognised that the national legislative process can be a lengthy one and that not all instruments can be adopted at the same time. Also, because the process of moving toward a nuclear program can be expected to take several years, it is reasonable to ask whether adoption of some instruments can be deferred until some later time. In essence, the best option would be to take action as early as reasonable, given other national priorities.

Two central instruments warranting early action are the Convention on Nuclear Safety and the Convention on the Physical Protection of Nuclear Material. Many of the provisions of both the CNS and CPPNM apply to the early stages of a nuclear programme or are necessary for security involving nuclear material. Early adherence to the CNS would allow Malaysia to participate in relevant review meetings at the beginning phases of its programme, something that would be helpful in keeping Malaysia up to date with developments in the field. Although it might seem that adherence to the Joint Convention on the Safe Management of Spent Fuel and Nuclear Waste might be deferred to a later time when Malaysia has generated quantities of waste, it is important to note that many of the steps needed for an effective waste management programme need to be taken at an early stage (for example, financing arrangements). The "bottom line" therefore, is that timing of adherence to relevant instruments is a matter for Malaysian authorities to decide given the overall situation in Malaysia.

The process of adopting an international instrument and implementing the obligations that accompany becoming a party to the instrument generally requires that a state undertake a series of steps to incorporate into their national law requirements that ensure that the obligations can be met. The requirements do not themselves have to incorporate each of the provisions set forth in the instrument, but they must ensure that the necessary legislative, regulatory and institutional infrastructure has been or can be established, implemented, maintained and sustained to adequately and effectively meet the obligations.

Knowing where a requirement might best fit within a State's legal system facilitates examination of whether the State's existing legislative, regulatory and institutional infrastructure is sufficient to cover incorporation and implementation of the new obligations. In that regard, in 2006 a "complete legal framework against terrorism" was included in UNODC's *Guide for the Legislative Incorporation and Implementation of the Universal Anti-Terrorism Instruments* (see pp. 10–11).[5]

c) *Human Capital Development*

To support the development of these nuclear technology applications, relevant academic programmes must be initiated at local universities. In addition to the formal university education, training and fellowship opportunities, such as to attend scientific visits, short term courses, fellowship attachments, workshops, seminars and conferences in other countries under international and bilateral cooperation are also used for the human capital development. For international and regional cooperation, the major avenues for human capital development are through the regular IAEA TACP, as well as Extra-budgetary Programme (EBP), Asian Nuclear Safety Network (ANSN), Asian Network for Nuclear Education and Training (ANENT), and RCA, which are all under the auspices the IAEA, besides the FNCA, funded by the FNCA member countries, independently from the IAEA.

To achieve the objective for Malaysia to be fully developed by 2020, national policies on education and training should focus more on science and technology, with a target ratio of 60:40 of science to arts students in secondary and tertiary education, while also enhancing scientific R&D in local universities. This should provide a stronger basis for human capital development towards promoting both energy and non-energy applications of nuclear science and technology. For this, international, regional and bilateral cooperation could be used to develop specific nuclear education programmes in local institutions. However, these programmes should address all levels of human capital development for all stages and activities that are planned for both energy and

---

[5] The Guide can be downloaded in PDF format from the UNODC website: http://www.unodc.org/documents/terrorism/Publications/Guide_Legislative_Incorporation_Implementation/English.pdf.

non-energy applications of nuclear science and technology in the country.

d) *Public Information and Stakeholder Relations*

Of critical importance to the promotion of both energy and non-energy applications of nuclear science and technology is an effective and focused public awareness programme, based on a timely dissemination of objective, accurate and relevant information. With the advent of the internet and greater digitisation of all types of information, various web-based textual, imagery, audio and video information are increasingly and creatively used. To ensure the effectiveness of the public information and awareness programme, public opinion surveys should be periodically undertaken to determine the key issues of public concern on both the energy and non-energy applications of nuclear technology, with the first survey serving as a baseline. The benefits and impacts of nuclear energy, and nuclear, radiation and related technology applications on the country, including on the well-being and safety of the public, workers and the environment, and competitiveness of local industries in various sectors in terms of production costs, and other such related matters, also need to be conveyed.

Conclusion

Although each of the international instruments and guidance documents contains specific provisions that Malaysia would need to implement, it is possible to generally outline the advantages and disadvantages of adherence to those which have been identified as essential or important for a nuclear power programme. It is submitted that the advantages for Malaysia strongly outweigh the disadvantages. By confirming that Malaysia will apply internationally recognised standards and procedures for ensuring the safety, security, safeguards and liability aspects of its program, adhering to the instruments will enable broad cooperation and assistance from foreign vendors and suppliers and international organisations. The instruments provide a useful template for measures Malaysia will need to implement in its national laws and regulations to ensure effective implementation of measures to protect the public health, safety and security of Malaysia's citizens. Participation in the review meetings of several of the instruments will keep Malaysian authorities current with the latest international consensus on measures needed to ensure safety, security and safeguards over any NPPs operating in Malaysia.

Chapter 5

POLICY CONSIDERATIONS FOR SINGAPORE'S NUCLEAR ENERGY INVOLVEMENTS

John Bauly[1]

Introduction

In this chapter we consider the backgrounds to nuclear policy and governance and how these suggest some policy actions for Singapore. These are summarised at the end of the chapter and are from the viewpoints of the writer.

Presently, as of 2014, approximately six per cent of the world's energy, including around 13% of electricity, is generated by nuclear power plants of a size up to 6000 megawatts (MW) and using one or more individual reactors of around 500–1400MW. Since Fukushima, the anti-nuclear argument has strengthened to express more forcibly that nuclear power is fundamentally flawed — primarily because it is seen as not safe *enough*. Other objections come from concerns about relatively high capital costs, very large unacceptable overruns of construction cost and leadtimes, undue outages, weapons proliferation fears, the extreme due diligence and security needed — but very hard to achieve, the lack of

---

[1] Dr John Bauly is a University Senior Fellow of NUS covering all aspects of NPD, i.e. New Product Development — as a Business Process including Management of Technology (MOT), marketing, design, etc.

effective nuclear governance as well as negative views from the media and public. Also notably several countries, like Germany which were previously very keen on nuclear power have, after careful reconsideration, now opted out of it.

Therefore, why should Singapore, or any country for that matter, want to try to cope with the difficulties of nuclear power, in the face of strong expert opinion from countries like Germany, Austria and Switzerland, Italy and others, that nuclear power can never be safe *enough* due to reasons of *unavoidable* imperfections in the technology and governance, and when gas, coal, oil, and the renewables seem so much easier to deploy, manage and continuously improve, and particularly when there is some hope that carbon capture technology might actually be viable? The answer is that there is strong pro-nuclear argument too, of course, plus some feeling of country national pride and technology stimulation in having a successful nuclear energy sector in one's energy mix.

Pro-nuclear opinion is that nuclear power is safe, clean, cost-effective, and should be expanded fast, world-wide, to combat the "Business as Usual", BAU, expected rapid rise in carbon emissions. From the IPPC work on their "Fifth Assessment Report", (IPCC, 2013),[2] these CO_2 emissions will likely result in the CO_2 atmospheric concentrations reaching the alarming levels of over 450 parts-per-million (ppm) of CO_2 by 2040. It is sobering to note that if these atmospheric concentrations happened to be carbon *mon*oxide rather than *di*oxide, all human life would disappear. Fortunately, because the combustion technologies ensure burning the fuels to almost all CO_2 we do not have to worry about that. A major way to combat these BAU dire CO_2 predictions is by substantially increasing the deployment of nuclear power. If deployed sufficiently, it ought to be possible to see a substantial decrease in the CO_2 world-wide emissions after say 2040. That would at least slow down the

---

[2] IPPC (Intergovernmental Panel on Climate Change), (2013). Climate Change 2013, The Physical Science Basis, Summary for Policymakers, (https://www.ipcc.ch/pdf/assessment-report/ar5/wg1/WGIAR5_SPM_brochure_en.pdf; accessed 20 April 2016).

rate of increase of the CO_2 atmospheric concentrations, and might even reduce it. No doubt the IPPC are working on those scenarios. Also by becoming less dependent on fossil fuels from other countries, the security of energy supplies becomes easier to achieve. Regarding "safety", the number of lives lost per megawatt hour (MW-h) for coal power is, according to the nuclear power enthusiasts, several times that for nuclear power. The unique challenging problem with nuclear accidents is their individual violence, geographical reach, and the possible harm to the neighbouring countries.

Singapore's Position

Singapore, an island of around 50 kilometres (km) by 25 km, with a population of around five million, is located at the southern tip of West Malaysia, and separated from it by a strip of sea just two to four km wide. Its second nearest large neighbour is the Indonesian Sumatera mainland around 100 km away. But Indonesia consists of many islands. A couple of these are Batam and Bintam, both comparable in size to Singapore, and just some 25–50 km to the south.

Singapore is arguable too small, and too highly populated, to ever host a nuclear power plant of conventional size. It might in the years ahead, consider deploying some very small modular reactors (SMRs), but only after they have been proven safe enough, and with their worst case scenario accident impacts known beyond doubt to be tolerable. SMR vendors will likely approach Singapore as well as the other ASEAN countries. Singapore and the other ASEAN countries should request from these vendors the worst-case accident impacts and the ASEAN countries usefully ought to share the results.

Although Singapore has little, if any, interest in nuclear power for itself, its neighbours Malaysia and Indonesia appear committed to go ahead with nuclear as soon as they can put in place the required governance, legislation, and technical abilities. Thus the question is what stance should Singapore take? On the one extreme it could adopt a "do nothing", "wait and see" approach, or at the other extreme it could seek to become highly proactive, and seek to influence its neighbours towards the highest standards of safety and governance and its own absolute immunity from the effects of any nuclear accidents particularly in Malaysia and Indonesia.

The present nuclear policy of Singapore results largely from a pre-feasibility study commissioned in 2010 by the Singapore Government. Following that, they issued a short explanation in October 2012 that "…. we prefer to wait for technology and safety to improve further before reconsidering our options"and "Singapore should play an active role in global and regional coop-eration on nuclear safety." Singapore will not want to be compro-mised, or lose opportunities, by "too little — too late", so prudently they will be already thinking about adding some specifics into the above policy. Hopefully this chapter will be helpful to them.

ASEAN and Nuclear Power

Singapore, Indonesia, and Malaysia, along with the Philippines, and Thailand, formed ASEAN in 1967, which has grown to include Brunei, Myanmar, Cambodia, Laos, and Vietnam. Similar to the EU, its aims include improving economic growth, social well-being, cultural development, protection of regional peace and sta-bility, and opportunities for member countries to discuss differences peacefully, bilaterally or multilaterally. ASEAN has a population of approximately 600 million people, which is around nine per cent of the world's population. By 2012, its combined GDP had grown to more than USD 2.2 trillion. If ASEAN were a single country, it would rank as the eighth largest economy in the world.

Vietnam in late 2012 started constructing two reactors of Russian design, and could have around 14 plants by 2030. No other nuclear plant constructions are in progress in ASEAN. Some other countries have carried out planning studies, but some were revised following Fukushima. The Philippines built the first nuclear power plant in ASEAN in 1985, but it was never commissioned, mainly because of safety defects, fears because it was built near major earthquake fault lines, and the anxiety from the 1986 Chernobyl disaster. Thailand has a research reactor and says it might have two plants operational by 2027. Indonesia has three research reactors but no firm build plans just now. Malaysia has a research reactor and says it might start a build in 2021.

It is not a good sign that the Fukushima disaster threw some countries' nuclear plans into disarray. That suggests, as a "hard truth", that the plans were naive, and the planning process flawed. Such extreme disasters are, at least with the larger plant, obviously unavoidable from time to time. That was clear already from Three Mile Island and Chernobyl. So the possibility should already have been recognised in the plans. Possibly, it was not realised that extreme natural or man-made disaster events are much more likely to occur than is predicted by the over-used Gaussian statistical distribution, because the real-life distributions has so-called "fat tails".

ASEAN has some similarities to the EU which was formed in 1957. The populations are roughly about the same, but ASEAN's GDP per capita is around just 15% of that of the EU, but growing fast — unlike the levelling off in the EU. The EU has developed its own "federal sovereignty" — so that all its countries citizens are also EU citizens too. The EU has a legal framework with laws and courts of justice. EU laws sometimes override country laws for individual human rights cases as well as for country laws defining broad policies. There are several EU regulations regarding energy, emissions like CO_2, renewables, nuclear safety, fuel cycles and safeguards. Such regulations are proposed and drafted by the EU Commission, and decided upon by the EU Parliament of 760 Members elected by public elections in all the member countries. The EU Council defines the policy agendas, for example it would have said some 15 years ago to the EU Commission: "We need to develop an overall energy policy, covering emissions, fuel mix, energy trading, and so on ...." The resulting EU regulations do provide some nuclear governance for the EU countries — including some governance regarding the risks to one country of nuclear plant located on its neighbour`s territory.

Some of these EU regulations relating to "nuclear" may be very useful for ASEAN reference along with those from the IAEA and the USA. A study of their usefulness to ASEAN would surely be worthwhile: rather than re-invent the wheel it would be wiser for ASEAN to see how it could adopt those existing regulations, and also practices, for their own use.

The ASEAN block was not created with the ambition of following the EU model. Whether or not it does eventually move in that direction in the decades ahead is very unclear. Just now one of its principles is to avoid bringing ASEAN group pressure to bear on other individual members. Of course individual ASEAN countries are still free to press arguments on to a colleague country if it is so minded. But that is little different to the bilateral discussions that any two countries can have anyway. However, an ASEAN Network of Nuclear Regulatory Bodies or Relevant Authorities, ASEANTOM, has recently been set up and this might be able to help, or even take a lead position regarding nuclear governance.

The "Governance Gaps"

By "nuclear governance" we mean that countries, for their nuclear power industry, firstly have in place a set of regulations and practices which are "complete and correct" as verified by peer review — preferably international peer review, and secondly have effective means of policing that the regulations are followed. Thirdly, the policing at national level ideally needs to be overseen at international level. This latter may be contentious, but the need for it is shown by the Fukushima disaster. Arguably, that awful disaster would have been avoided if the Fukushima authorities had acted reasonably to upgrade their Tsunami defences, following the IAEA advice given to them about the extreme event of the massive Indonesian Tsunami of 2004, seven years before Fukushima.

Clearly, there appears to be "governance gaps" of effective means to make sure that recommendations from agencies like the IAEA are acted upon with due diligence.

This disappointing lack of action in the case of Fukushima supports pro- and anti-nuclear lobbies alike. "Pros" say that we now know that some countries, if indeed not all, need an international level of supervisory policing regarding safety. Others argue that, actually, international supervisory policing is needed for all countries just like that which is somewhat achieved for the international airline businesses. The need for overseeing at international level is verified by the Chernobyl disaster when neighbouring countries severely suffered from the Ukraine mishap. Obviously, then, nuclear governance has to specifically look after the interests of neighbours' safety too. At present it does not do so fully, so this is another one of the governance gaps.

"Antis" just use Fukushima to demonstrate that indeed nuclear is and always will be "playing with fire", and simply too dangerous.

One of the difficulties about the nuclear safety issue is that the nuclear plant vendors and developers are addressing it with an approach of making the chances of major nuclear accident impacts "very unlikely" — by higher safety margins, redundancy designs, additional levels of controls, and so on. This tendency was followed by the Japanese after Fukushima, on the lines of "see how we are now going to make such accidents in future very unlikely — with various clever technical improvements…" We do not notice much addressing by the Japanese of the failure in governance, nor about the selection of designs in the future which have much lower worst case accident impacts. We thus see the trend of offsetting high possible "impact" by lower "chance". The tendency of not wanting to define the worst case accident impacts is one of the governance gaps. During the 2013 Singapore international Energy Week Roundtable on Nuclear Governance, the speaker from Malaysia said, quite rightly: "We have to be prepared for the worst" (i.e. accident impact). That cannot be done if the worst case accident impact is unknown. Neither is it acceptable, or ethical, to just say "such accidents are very unlikely".

If the governance gaps are not attended to, as they become more obvious, they will no doubt undermine any realistic chance of a nuclear renaissance.

It should be remembered that the opinion of nuclear power worldwide tends to be polarised. The Fukushima disaster was the catalyst which made the Germans, for example, turn 180 degrees against nuclear power. As the governance gaps now becomes more recognised worldwide, inevitably, some now "pro" opinion formers will turn 180 degrees too and join the "anti" camp.

Country Culture

Besides nuclear governance, there is the sensitive issue of "culture". For nuclear safety, an extremely high total quality of due diligence, personal and institutional integrity, and review processes is absolutely vital. This particularly recognises that the majority of accidents are caused by human error or misbehaviour. Even for cases where a system failure starts an accident event sequence, human failure to apply available remedial action can often lead to further escalations. The hiding of mistakes or problems, not promptly asking for help, carelessness, dishonesty, and inadequate knowledge and supervision, are entirely unacceptable. So cultures where workers, including professionals, are liable to behave badly like that, or where the supervisors "shoot the messengers", or demand "good stories" to please their management, result in dangerous exposures. High total quality behaviour is particularly demanded in other life hazardous areas like the worldwide airline businesses, and the medical profession, as examples. But it is only achieved after some years of continuous improvement, and then only if it is upfront recognised as a critical success factor. It is hard to achieve in any country, and more so if the levels of education, industrialisation, services, and the behavioural examples set by individuals in government and public service are under par. Nevertheless, arguably with very careful selection of staff at all levels, excellent orientation and training, nuclear plant staff, and the governance management in any country could be "world class". Part of the job of effective governance should be to ensure this, verifying it by international peer review.

Singapore's Expectations

Within ASEAN, and indeed in the world, Singapore stands out as different. With virtually no natural resources, a high population density, the GDP per capita is at around USD 60,000 and its level of education and life expectancy are amongst the highest in the world. The GDP per capital for its nearest ASEAN neighbours is around USD 17,000 for Malaysia and USD 9,000 for Indonesia, but happily both are growing fast. These countries, along with most other ASEAN countries are therefore still in the developing phase of their fast growth trajectories. That might mean that extra special efforts are likely required to achieve the needed "world class" level for staffing and managing their nuclear plant and its governance. Singapore would want to be assured of that before any nuclear plant was placed in Indonesia or Malaysia within low level harm reach of Singapore. This would just be "reciprocally fair" in that Singapore would expect both Indonesia and Malaysia to want just the same assurances from Singapore if it ever did decide on having nuclear plant on its soil. Thus a dialogue on those lines with Malaysia and Indonesia, hopefully leading to agreements, might usefully be progressed. That would be part of the policy movements one might expect of Singapore. Such discussions ought not to be too difficult nowadays.

Singapore's Outlook Towards Malaysia, Indonesia, and ASEAN

In Singapore some opinion has been expressed during debate between students and academics that Singapore should try to establish itself as having a lead position in ASEAN regarding nuclear power knowledge and also that Singapore ought to acquire a research reactor mainly to heighten its standing in the subject. For sure, these competitive type policies are not the best. Other parties with an obvious knowledge decrement would be too defensive in negotiation and become obstructive. Best policies are about mutual help, learning, and "win-win". Regarding research reactors, one might be useful for Singapore, but only if the use of it was of real commercial value, the research results usable, and if the students acquiring training from it could gain employment afterwards. A research reactor in Singapore was looked at before, and not proceeded. Here we again note that several ASEAN countries do have research reactors, including Malaysia and Indonesia. One area of useful mutual "win-win" learning ought to be collaborative reviews of the research reactor programmes and the commercial outputs of isotopes. Could Singapore be a good customer? Could there be some collaborative training and research projects? How is security handled? In negotiations with nuclear plant vendors, the IAEA, and other parties, one does not want to be in position of low information imbalance. That is a good driver for Singapore, Malaysia, Indonesia and other ASEAN partners to set up together some focussed structured study programmes to learn more about the subjects. The people being so briefed could include a few from government departments and a few from university. It would be done by self-learning, attending a key international conferences, these being followed by written evaluations, perhaps all facilitated by two or three tutors from home or abroad. The subjects would include about accident impacts: how they depend on type, size and design parameters, severities and emergency services, and clean up; and include SMRs. Other areas would be about the degree of certainty of costs and lead times, fuel reprocessing, training of staff, governance arrangements, and related areas. The studies should also include reviews that the processes for determining overall and detailed policies are robust and realistic.

In recent years, the concept of "preferred safe designs" has been floated, and that these might be based on the small modular reactor (SMR) concept — either as single standalone plants, or as larger plants made up of several SMRs. That concept might well interest the ASEAN countries, even as a game changer for them, i.e. to switch to SMRs rather than continue with the much more hazardous commitment to the traditional large designs. A collaborative study of this within ASEAN would surely be appropriate.

Suggested Singapore Policy Actions

The Singapore policy actions we suggest in this chapter are firstly consistent with the conclusions of pre-feasibility study on nuclear energy as reported by the Singapore Ministry of Trade and Industry in October 2012, and secondly also consistent with the Research and Education Programme in Nuclear Safety, Science and Engineering announced to be the Singapore National Research Foundation on 23 April 2014. However, our policy action suggestions address more specifics as well as additional items and are summarised as follows:

(i) Set up "win-win" collaboration with Malaysia and Indonesia centred around the commercial, research and training activities of their research reactors. This collaboration might be bilateral or multilateral. Notice that although Singapore does not have a research reactor, it has plenty to offer for information exchange. NUS have a strong nuclear physics activity with world leading positions for example in proton beam technology. Singapore also has other world leading advanced technology, including targeted radiation therapy like PRRT (Peptide Receptor Radionuclide Therapy) for some cancers, now operational at Singapore General Hospital, as well as advanced nuclear imaging.

(ii) Develop (i) to include other interested ASEAN countries.

(iii) Develop (i) and (ii) to run some "teach-in" joint study programmes on nuclear technology, plant management, governance, economics, strategic planning, accident impacts and incident management.

(iv) Lobby those ASEAN countries considering nuclear power, to define the maximum acceptable worst-case accident impacts, depending on location, even when such accidents are said to be "very unlikely" — and share their results.

 (v) Develop the above to establish governance agreements, at least with Malaysia and Indonesia, that the location of any nuclear plant must consider the possibility of undue harm to its neighbours as well as to themselves.

(vi) Lobby for a collaborative study on the pros and cons of SMRs to be carried out i.e. as an ASEAN project.

(vii) Lobby the IAEA to encourage, or even require, nuclear plant vendors and developers to determine and report their assessments of the worst-case accident impacts of their plant — for SMRs as well as large plant.

(viii) Lobby the IAEA that this data (vii) should then be analysed by relevant technical institutes, probably abroad, helped by the IAEA, to study how those impacts depend on the type and size of plant, and furthermore to determine if "preferred safe designs" are thus suggested.

(ix) Agree with ASEAN countries to request the information of (vii) from all vendors they deal with, and share the results with each other.

 (x) Lobby ASEAN members to jointly study how existing EU and USA nuclear regulations and practices might be adopted usefully be ASEAN countries.

(xi) Determine to what extent ASEANTOM already covers some of the above, or could do so.

Some of the above could run concurrently. Some might be handled via existing ASEAN collaboration arrangements.

(xii) Presently a number of Singapore Government Agencies are concerned with the different facets of nuclear technology and power. These include National Research Foundation, NRF; Ministry of Trade and Industry, MTI; and the National Security Coordination Secretariat, NSCS. There might be some advantage of focus if a small "nuclear directorate" office was set up to overall coordinate, lead, and suggest needed initiatives.

References

ASEAN Secretariat (June 2009). ASEAN Socio-Cultural Community Blueprint. Jakarta: ASEAN Secretariat, (http://www.asean.org/archive/5187-19.pdf; accessed date 20 April 2016).

Bauly, John (April 2013). Why Germany abandons nuclear power — and what does that mean for Asia? *Asia Europe Energy Policy Research Network*, (http://www.aeeprn.com/commentaries-item/2014/03/21/why-germany-decided-to-abandon-nuclear-power-and-what-does-that-mean-to-asia-; accessed 20 April 2016).

Bauly, John (24 October 2013). Small Modular Nuclear Reactors: Game Changer? Energy Studies Institute, National University of Singapore, (http://www.esi.nus.edu.sg/docs/default-source/event/smrs-esi-seminar.pdf?sfvrsn=0; accessed 22 April 2016).

Energy Outlook 2013. Washington DC: US Energy Administration, (http://www.eia.gov/forecasts/ieo/pdf/0484(2013).pdf; accessed 29 April 2016).

Govt to conduct feasibility study on building nuclear plant: Mah (7 July 2014). *New Straits Times Online*, (http://www.nst.com.my/node/10529; accessed 18 August 2014).

International Enterprise (IE) (2012-2013). 30 Years of Globalising Singapore, Annual Report 2012/2013, (http://www.iesingapore.gov.sg/annual-reports/2012-13/; accessed 23 April 2016).

International Atomic Energy Agency (IAEA) (4–7 September 2012). IAEA Report on protection against extreme earthquakes and tsunamis in the light of the accident at the Fukushima Daiichi nuclear power plant. *International Experts Meeting Vienna*, (https://www.iaea.org/sites/default/files/protection040912.pdf; accessed 24 April 2016).

Kousky, Carolyn and Cooke, Roger (May 2010). Adapting to Extreme Events: Managing Fat Tails. *Resources for the Future*, Issue Brief 10–12, (http://www.rff.org/RFF/Documents/RFF-IB-10-12.pdf; accessed 24 April 2016).

Ministry of Trade and Industry (MTI) (15 October 2012). Factsheet: Nuclear Energy Pre-feasibility Study. Singapore: MTI, (http://www.mti.gov.sg/NewsRoom/Documents/Pre-FS%20factsheet.pdf; accessed 26 April 2016).

Ministry of Trade and Industry (MTI) (2013). Yearbook of Statistics Singapore, 2013. Singapore: Department of Statistics, Ministry of Trade and Industry.

National Research Foundation, Prime Minister's Office (23 April 2014). Establishment of Research and Education Programme in Nuclear Safety, Science and Engineering, Singapore: National Research Foundation, Prime Minister's Office, (http://www.nrf.gov.sg/docs/default-source/Press-Releases/20140423_nsrep-press-release-(final).pdf?sfvrsn=2: accessed 28 April 2016).

Nian, Victor and Bauly, John (30 May – 2 June 2014). Nuclear Power Developments: Could Small Modular Reactor Power Plants be a "Game Changer"? – The ASEAN Perspective. *International Conference on Applied Energy.* Taipei, Taiwan: ICAE 2014, (http://www.science-direct.com/science/article/pii/S1876610214027258; accessed 28 April 2016).

Phruksarojanakun, Phipat (November 11–12, 2013). ASEANTOM: ASEAN Network of Regulatory Bodies on Atomic Energy. CSCAP Nuclear Energy Experts Group, (http://csis.org/files/attachments/131111_Session%205_Phiphat.pdf; accessed 28 April 2016).

Putra, Nur Azha (May 2014). Singapore's Role in International Nuclear Safety and Security Cooperation, ERIA Final Report.

Singapore General Hospital (29 December 2014). Nuclear Medicine and PET Overview, (http://www.sgh.com.sg/Clinical-Departments-Centers/Nuclear-Medicine-PET/Pages/overview.aspx; accessed 29 April 2016).

Subki, M Hadid (18–20 June 2013). Update on IAEA Programme on SMR Technology, Development and Deployment. *IAEA Meeting of the TWGs of Advanced Technologies for LWRs and HWRs,* (http://www.iaea.org/NuclearPower/Downloadable/Meetings/2013/2013-06-18-06-20-TWG-NPTD/33-iaea-smr.pdf; accessed 29 April 2016).

Chapter 6

HUMAN RESOURCES AND CAPACITY BUILDING: ISSUES AND CHALLENGES FOR VIETNAM'S NUCLEAR ENERGY PROGRAMME

Ton Nu Thi Ninh[1]

Introduction

Among the countries of Southeast Asia, Vietnam has quite early on (since the 1980s) chosen a nuclear option for its planned energy mix over time. This choice has been explicitly and consistently reiterated, including after the Fukushima disaster. Today Vietnam is at the forefront of nuclear energy development in Southeast Asia with 2020 (2025) as the planned start year of commercial operation for the first nuclear power unit. Within Vietnam, however, although the nuclear option is not the object of much or broad debate, a number of scientists and experts have raised the question of human resources and capacity building preparedness as perhaps the top challenge arising from the nuclear energy option. It is therefore important to look into the status of the issue from a Vietnamese perspective.

---

[1] Ambassador Ton-nu-thi Ninh is a member of Vietnam's law-making body, the National Assembly.

Vietnam's Nuclear Energy Planning and Development

Vietnam's interest in and decision to take up a nuclear energy option is not recent, dating back to the early 1980s (1980–86) when a project for the Development of a Nuclear Power Plant (NPP) in Vietnam was presented by Vietnam's Atomic Energy Institute. This project won the support of the Prime Minister and initial talks with the Soviet Union were conducted for the transfer of a 440 megawatt (MW) NPP.

There was a twenty year hiatus (1986–2006) during which the country focused on carrying out the so-called "Đổi mới" (reform and renewal) or economic restructuring. But the nuclear energy option was consistently reaffirmed and even given added foundation and scope with the adoption in 2006 of the Prime Minister's Decision on the Strategy for Peaceful Uses of Atomic Energy up to 2020. This policy choice was further strengthened with the adoption in June 2008 by Vietnam's legislative body — the National Assembly — of the Atomic Energy Law, thus sealing a consensus between the executive and the legislative on the nuclear option.

Since then the roadmap towards Vietnam's first (NPP) has been gradually implemented. In November 2009, the National Assembly passed the Resolution on the Ninh Thuan NPP. The foreign partners for the Ninh Thuan NPPs have been chosen: Russia for Ninh Thuan 1 and Japan for Ninh Thuan 2. The choice of these partners can, to some extent, be explained by political as well as financial factors: Vietnam's long-standing relationship with Russia — including in relation to the issue of nuclear energy dating back to 1986 as was mentioned earlier; Japan being Vietnam's top overseas development assistance donor as well as the country's top foreign direct investment partner; and both countries realistically joining a financing setup to their submission. No decision has yet been taken on the choice of technology but Vietnam is pleased to know that the Japanese partner intends to subcontract part of the Ninh Thuan 2 NPP project delivery to France, which has a significant and sustained track record in the civilian nuclear energy industry.

In 2010, Vietnam's Prime Minister approved a master plan for the Peaceful Development and Use of Atomic Energy up to 2020. Detailed planning developed between 2010 and 2012 covering multiple areas such as agriculture, medicine, industry, geological science and environment, nuclear power development, HR development, disposal site, capacity building for security, and capacity building for research & development.

Nuclear energy planning in the context of Vietnam's overall power generation planning is as follows: in 2010 Vietnam's power generation amounted to 97.25 billion kilowatt hours (Kwh). According to Master Planning VII for power generation approved in 2011, by 2020 Vietnam's power output would grow threefold to 330 billion Kwh and by 2030 a further twofold to 695 billion Kwh. The amount projected for nuclear energy would be 2.1 per cent (2020) and 10.1 per cent (2030).

In sum, Vietnam's nuclear energy choice goes back several decades and practical planning started some 10 years ago.

Post-Fukushima, the nuclear energy option has been reaffirmed taking into account lessons from Fukushima that were shared by the International Atomic Agency (IAEA): some modifications to the location (place & height) of the planned Ninh Thuan 1 NPP, safety and being up-to-date with the choice of technology has been emphasised, and the HR training planning as well as the legal framework have been revisited (the 2008 Atomic Energy Law being reviewed for amendment; Prime Minister approval in February 2013 of the list of legal documents to be promulgated up to 2020),

Table 1: Master Planning VII for electricity development for the period of 2011–2020 with consideration up to 2030.

	Thermal	Hydro	Gas	Renewable	Nuclear	Imported
2020	46.8%	19.6%	24%	4.5%	**2.1%**	3%
2030	56.4%	9.3%	14.4%	6%	**10.1%**	3.8%

Source: Nuclear Power Programme in Vietnam, Hoang Anh Tuan, PhD., Director General, Vietnam Atomic Energy Agency — October 2013.

enhancing of state agencies and other actors' institutional capacity.

One should add that in Vietnam no such domestic political pressure on the nuclear issue exists as in Europe or Japan, only some degree of scrutiny by scientists and experts.

What has Been Done so Far in Terms of Capacity Building and Human Resources Training?

Reflecting the official commitment of the Vietnamese government a full governmental steering structure has been put in place involving both central and local authorities as well as the nuclear power operator — Electricity of Vietnam (EVN).

1. VARANS: Vietnam Atomic Research Nuclear Safety.
2. VAEA: Vietnam Atomic Energy Agency.
3. VINATOM: Vietnam Atomic Research Institute.

As the planning on peaceful utilisation of atomic energy up to 2020 shows, a whole set of practical, concrete implementation measures are being diligently charted. These measures put specific emphasis on safety and security, human resources and public communication reflecting a measure of sensitivity to the post-Fukushima heightened concern — at least internationally — to nuclear safety and to the need for public acceptance of nuclear energy.

In 2010, the National Project for the Training and Development of Human Resources for Atomic Energy was approved. In 2011, the government launched the Project on the Development of Measures to guarantee Safety & Security in the Field of Atomic Energy. In 2012, the National Project on Public Information up to 2020 was initiated and by May 2013, the National Nuclear Safety Council was established. In view of the Fukushima incident Vietnam is setting up a National Preparedness and Responses plan involving three levels: nationwide; local (Ninh Thuan province); the NPP itself. Currently work has started on the nationwide framework.

More specifically with regard to HR training planning, the master plan for "Training and HR Development in the Field of Nuclear Energy" was approved by the Prime Minister in August 2010. A national steering committee for the implementation of the HR master plan was set up and led by the Deputy Prime Minister and has held five review meetings so far. Components of the HR master plan consist of:

• HR development to be one step ahead of infrastructure development and operational development;
• Need to ensure safety and security;
• Due attention to policy planning and legal training for management;
• Attracting overseas Vietnamese expertise and engaging proactively in international cooperation.

In terms of HR training planning, five universities[2] and one training centre under the Institute for Nuclear Energy (Ministry of Science and Technology) were selected to carry out training. Training targets set for the Ninh Thuan 1 NPP by 2020 include more than 2,400 engineers (200 trained overseas) and more than 350 Masters and PhDs in Nuclear Power (150 trained overseas).

More specifically for R&D and safety & security, it is expected that 650 engineers (150 overseas) and 250 Masters and PhDs (100 overseas) will have been trained by 2020 while 100 Masters and PhDs are earmarked to be trained as nuclear energy trainers. 500 administrators, managers and scientists should also be sent on observation & study tours in relevant countries. Another task consists of reviewing and modifying related curricula and syllabi towards being up-to-date, closer linkages between theory and practice as well as between academia and Research and Development.

---

[2]University of Natural Sciences — VNU (Ha Noi); Polytechnic University (Ha Noi); University of Natural Sciences — VNU (Ho Chi Minh City); Dalat University (Dalat); Electric Power University (EPU, Ha Noi).

A number of practical steps have been taken since 2010:

• Between 2005 & June 2012 Vietnam Electricity Corporation (VEC) dispatched 196 staff on short training courses abroad (Japan, Korea, France, Hungary, etc.) and 29 students on long-term training at the Moscow Power Engineering Institute and two to Grenoble University in France. VEC started training at its own university (EPU) in 2010 with a yearly intake of 50 students.
• It is expected that each of the five other institutions assigned will take in 30 students every year from 2014.
• Overseas training has started with Russia (238 so far), Hungary (short term courses for 496 faculty members) and Japan.
• In tandem supporting measures have been taken, with MOUs signed with ROSATOM of Russia in March 2010 for training of Vietnamese students and in October 2011 for the establishment of an Information Center on Nuclear Energy in Ha Noi financed by ROSATOM.

On 18 September 2012, an MOU was signed between Vietnam and IAEA on cooperation for HR training in atomic energy including consultancy related to curriculum, syllabi, equipment, search for appropriate academics and scientists, and quality assurance regarding Vietnam's atomic energy training programs.

This was followed in September 2013 by the signing of an agreement on atomic energy training with Hungary; and another with Japan. Roughly USD 150 million has been mobilised to serve the HR master plan.

• From all of the above, the following can be asserted:
  — Vietnam's consistently reaffirmed nuclear energy policy over the past several decades, including post-Fukushima;
  — This translates into serious comprehensive and systematic planning taking into particular account the Fukushima incident;
  — Due importance is granted to the need for international guidance and cooperation in several aspects of nuclear

energy, including development of national nuclear power infrastructure and especially with regard to the safety, security and safeguards (3S) as well as capacity building and HR training. IAEA has conducted two INIR missions to Vietnam: one in 2009 (30 Nov–4 Dec) and one in 2012 (05–14 Dec).

— Most recently on the occasion of the Nuclear Power Asia 2014 Conference held in Ha Noi (20–22 January 2014), IAEA Director General Yukiya Amano paid a visit to Vietnam during which Vietnam's plans for a nuclear power plant were discussed.

- Vietnam's commitment to international standards translates into its active participation in most of the relevant international instruments:
  — 1982: Non-Proliferation Treaty (NPT)
  — 1987: Convention on Early Notification of a Nuclear Accident (accessed)
  — 1987: Convention on Assistance in Case of Nuclear Accident or Radiological Emergency (accessed)
  — 1989: Safeguards Agreement with IAEA
  — 1996: Bangkok Treaty on Southeast Asian Nuclear Weapon Free Zone (SEANWFZ)
  — 1996: Comprehensive Nuclear-Test-Ban Treaty (CTBT, signed, ratified 2006)
  — 2007: Additional Protocol (AP, signed, ratified 9/2012)
  — 2010: Convention on Nuclear Safety (CNS, accessed)
  — 2012: Convention of Physical Protection of Nuclear Materials and its Amendment (in force since 3 Nov 2012)

Moreover, consideration is being given to participation in the following instruments:

  — Joint Convention on the Safety of Spent Fuel Management and on the Safety of Radioactive Waste Management
  — Vienna Convention on Civil Liability for Nuclear Damage
  — International Convention for the Suppression of Acts of Nuclear Terrorism

Vietnam is also actively engaged in several relevant regional institutions e.g. the Forum for Nuclear Cooperation in Asia (FNCA); the Asia-Pacific Regional Cooperative Agreement (RCA); the Asia Nuclear Safety Network (ANSN); and the Asia-Pacific Safeguards Network (APSN).

Outstanding Issues and Challenges

Regardless of what has been described above, lingering scepticism remains among some experts, and in scientists' circles about the wisdom and feasibility for Vietnam of moving into nuclear energy at this stage. They argue that this option is better suited to developed countries; that if Vietnam's power utilisation was made less wasteful (power consumption growing more than twice as fast as the economy: 12.5 per cent to 5.3 per cent in 2012), the need for nuclear energy would be reduced. They stress further that post Fukushima nuclear energy is becoming expensive as much more complex safety measures have to be built in. In particular they believe it will be very hard if not impossible for Vietnam to meet the 2020 target in terms of appropriate capacity building and human resources.

In fact, their view is, now that the executive and legal framework is in place, the slower the process of implementation the better, as haste would result in unsafe preparations. Actually during his January 2014 visit IAEA's Director General urged Vietnam to avoid rushing to build a nuclear power plant and rather to focus on ensuring that this first NPP project is developed with utmost care. As a matter of fact the Minister of Science and Technology admitted that the construction of the first NPP scheduled to start this year (2014) may be delayed for two or three years in order to develop additional safety measures and thus Vietnam's first NPP might become operational only in 2025 (rather than in 2020 as planned earlier).

The executive institutional set-up is indeed full-fledged, with all related agencies involved. There remains, however, the issue of effective and timely interagency and central-local coordination especially in case of a disaster. Vietnam has proven experience in disaster preparedness and responses when it comes to typhoons and floods. However, Vietnam has no experience of nuclear-related incidents on its territory. As part of the national preparedness and responses plan, drills should be integrated in a suitable way.

In terms of qualified HR for the construction and operation of a NPP, Vietnam suffers from a clear deficit: most of the 568 staff (2012) working in the nuclear field is concentrated in the non-power nuclear industry such as the Vietnam Atomic Energy Agency. There is a consensus on considering capacity building and HR training for nuclear energy as the greatest challenge (both quantitative and qualitative) and top priority if Vietnam is to carry out successfully its nuclear energy master plan.

There is yet the issue of how to structure the HR training: some experts worry that too much emphasis — including quantitatively — is being put on theoretical academic training rather than on application. They think that what is needed is good engineers, more particularly those working in thermal or hydropower plants that are further trained in the specifics of nuclear energy.

It has been stressed that there exists a further challenge, specific to Vietnam, i.e. the deficit in "safety culture" in Vietnam: the Vietnamese can be viewed as at the opposite end of the Japanese for instance, who are known for their "safety mindset and culture". This may mean that HR manning the future NPP in Ninh Thuan might not follow strictly all safety rules and processes. Proper independent expert supervision and oversight might be necessary. Some even say that there is a need for building nuclear literacy for all ASEAN countries.

Finally, proper nuclear governance post-Fukushima requires responsible and timely public information to ensure public trust and support at all levels (national, provincial and local). In this respect, the fine line is to determine what information needs to be made public or available while avoiding causing unfounded counterproductive public anxiety.

In facing these challenges as a newcomer, Vietnam is also looking to fellow ASEAN countries to share experiences in preparing for the nuclear energy option as well as to link up with the ASEAN nuclear regulators network. In this context the Asia Pacific nuclear community concept put forth by the Asia Pacific Leadership Network for Nuclear Non-proliferation and Disarmament (APLN) may offer pertinent reference.

Bibliography

Electric Power University (2013). *Report on Human Resource Training for Nuclear Power Plant Projects in Ninh Thuan Province*. 27–29 August 2013. Ha Noi.

Government Office of Vietnam (2012). *Announcement on Deputy Prime Minister Nguyen Thien Nhan's Conclusion at the Second Meeting of The National Steering Committee on Training Human Resources in the Field of Atomic Energy*. 1 June 2012. Ha Noi.

Government Office of Vietnam (2013). *Announcement on Deputy Prime Minister Nguyen Thien Nhan's Conclusion at the Fourth Meeting of The National Steering Committee on Training Human Resources in the Field of Atomic Energy*. 17 January 2013. Ha Noi.

Government Office of Vietnam (2013). *Announcement on Deputy Prime Minister Nguyen Thien Nhan's Conclusion at the Fifth Meeting of The National Steering Committee on Training Human Resources in the Field of Atomic Energy*. 12 August 2013. Ha Noi.

Hoang, A. T. (2013). *Nuclear Power Programme in Vietnam*. Asia-Pacific Leadership Network (APLN) Meeting, 12 October 2013, Ho Chi Minh City.

Ministry of Education of Vietnam (2013). *Report on the Progress of "Training and Developing Human Resources in the Field of Atomic Energy" Scheme*. 26 August 2013. Ha Noi.

Nuclear plant to be delayed (21 January 2014). *Vietnam News*. Retrieved from: http://vietnamnews.vn/economy/250431/nuclear-plant-to-be-delayed.html; accessed 2 May 2016.

Prime Minister's Decision: Approval of Training and Developing Human Resources in the Field of Atomic Energy" Scheme. 18 August 2010. Ha Noi.

The National Steering Committee for Ninh Thuan Nuclear Power Plant Project (2013) Decision on Establishing Sub-Committee on Training and Communication. 29 May 2013. Ha Noi.

Chapter 7

SOCIOECONOMIC IMPACT OF NUCLEAR POWER

Hans Rogner[1]

Introduction

The socioeconomic impacts of nuclear power can be positive and negative and as always, involve trade-offs. Many of the impacts are plant and site specific, and subject to perception, which makes weighting and balancing of risks and benefits a somewhat subjective, tricky and controversial affair.

The construction time of a nuclear plant (up to ten years) exceeds the length of standard election cycles. A non-partisan national position, a proper policy and regulatory regime and stakeholder involvement on the inclusion of nuclear in a country's long-term energy mix, are essential for planning certainty, investor confidence and optimal resource allocation and use. The decision to build and operate nuclear power plants implies a national commitment of at least one century. Appreciating the socio-economic impacts of nuclear power, measured against those of its alternative

---

[1] Until his retirement from the International Atomic Energy Agency (IAEA) in 2012, Hans Rogner directed the International Atomic Energy Agency's programme on Capacity Building and Nuclear Knowledge Maintenance for Sustainable Energy Development.

electricity supply options, is therefore essential in a national debate on the potential role of nuclear power.

Economic Impacts

A nuclear power plant (NPP) represents an investment of between USD2 and USD8 billion,[2] depending on plant and site specific factors (new technology on a greenfield location or a standard plant on an existing site), geology and localisation factor) and whether finance and interest during construction are included.

Local

The employment effect of constructing a NPP averages 1,600 direct on-site labour and professional jobs for four to six years, with peak employment over 4,000. The economic impact results from that part of wages spent locally by workers on services (e.g. food, housing, health care, transportation, recreation, entertainment), generating another 1,000 to 1,400 indirect jobs around the plant site.

As NPPs are usually sited in areas with low population density, a large share of the additional employment migrates from other part of the region pushing demand for accommodation and services beyond the existing levels of supply causing inflationary pressures on prices and stress on local infrastructures. Construction can adversely affect local air quality (dust), increase both noise levels and traffic to and from the construction site.

The economic boom during plant construction is followed by contraction after construction completion when construction workers and construction support businesses move on. This often results in declining property values, surplus service capacities and lay-offs. The contraction is cushioned by the 400 to 700 permanent (non-construction) jobs created for the operation of an NPP and the additional indirect employment to support the plant and its work

---

[2] IEA/NEA (2010) reports median overnight cost of USD3,740 per kilowatt (kW) installed with a range of USD1,556 to USD5,863 per kW.

force for up to 60 years according to the Nuclear Energy Institute's (NEI) white paper on nuclear energy (NEI, 2013).[3] Generally, staff directly employed in the operation of nuclear facilities are full-time and long-term employees. The staff includes a high share of skilled professionals with income above national levels — hence overall positive economic impacts on the community and the region through direct and secondary spending (NEI 2013).[4]

NPP operation also provides long-term, stable tax revenue, generally resulting in above average local infrastructures and services (education, health care, transportation and recreational facilities). On the negative side, non-energy products from nuclear communities have been stigmatised in more distant markets due to the fear of possible radioactive contamination, e.g. of agricultural or fishery products.

Regional/National

NPP construction provides a substantial boost to suppliers of construction materials, commodities and manufacturers of plant components. The construction sector in general has a high economic multiplier effect: GBP1 spent on general construction in the UK results in an estimated GBP2.84 of total economic activity (Brookman, 2012).[5] The NEI (2013)[6] reports a macroeconomic multiplier of 1.87 specifically for NPP construction in the United States (US).

Indirect employment via service providers, material suppliers and component manufacturing can add another 1,000 to 2,000 jobs (and associated value added) depending on the local economy: the

---

[3] NEI (2013). Nuclear Energy's Economic Benefits — Current and Future. White Paper Washington DC 20004, USA: Nuclear Energy Institute, (http://www.nei.org/CorporateSite/media/filefolder/economicbenefitscurrentfuture.pdf?ext=.pdf; accessed 2 May 2016).
[4] Ibid.
[5] Brookman G. (2012). Bridging the gap — Backing the construction sector to generate jobs. Business environment directorate, London, UK: The CBI Company.
[6] NEI (2013), op.cit.

economic multiplier effect is considerably reduced for small open economies where most NPP components and many related services are imported.

The economic impact of nuclear power compared to alternatives depends on several largely location-dependent factors: (domestic sources or imports of fossil fuels, domestic vs. imported technology). Comparisons will also vary depending on how the comparison is structured: the boundaries (especially depth of indirect effects) of the assessments and the underlying assumptions, which vary greatly between different interest groups even for the same technologies.[7] Most studies assess the employment effects rather than the economic impact (employment, value added and price effects, etc.), which necessitates the use of detailed input–output models (rarely available) as well as an economic valuation of supply security, reliability, environmental performance or externalities.

Generating Costs

Figure 1 summarises the overnight investment cost (OC) data (without interest during construction (IDC)), from the Organisation for Economic Co-operation and Development's (OECD) study "Projected Costs of Generating Electricity — 2010 Update" (IEA/ NEA 2010).[8] The left panel shows an overlap and spread of specific investment costs for different generating technologies, reflecting varying local conditions, technology designs, regulatory and environmental constraints. The lower boundary represents the conditions in large developing countries such as China and India while higher prices reflect particularly challenging site conditions in OECD countries.

The right panel shows costs for different plant sizes, with nuclear plants more than twice as large as any other, highlighting

---

[7] See for example Table 2 of Wei M., Patadia S. and D. M. Kammen (2010). Putting renewables and energy efficiency to work: How many jobs can the clean energy industry generate in the US? *Energy Policy* **38** (2010), pp. 919–931.
[8] IEA/NEA (2010) Projected Costs of Generating Electricity, 2010 Edition. Paris, France: International Energy Agency and Nuclear Energy Agency of the OECD.

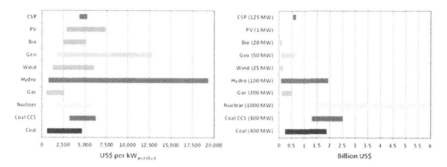

Figure 1: Overnight investment costs for 1,000 MW of different electricity generating technologies (left) and overnight capital costs for typical unit sizes (right panel).

Source: IEA/NEA 2010.

the importance of grid and market size in technology choice. Small grid sizes limit the integration of presently commercially available units of 1,000 MW or more. Smaller unit projects are easier to finance, especially for utilities with low capitalisation. In the future, the commercialisation of small and medium sized NPPs (100–600 MW) might ease financing and integration into smaller grids.

Long-run marginal generating costs accounting for OC, interest during construction (IDC), fuel, operating and maintenance costs, waste management and decommissioning costs are often used to rank investment alternatives. Figure 2 (overleaf) shows the ranges of levelised costs of electricity (LCOE) generation for real discount rates of five and ten per cent per year. Depending on local conditions, the LCOE range for nuclear power coincides with most competing technologies.

Price Stability

NPPs are expensive to build but cheap to operate. Nuclear generating costs are relatively stable and predictable over extended periods of time. Fuel costs (including spent fuel management costs) are only 20 to 25 per cent of the total cost compared to some 60 and

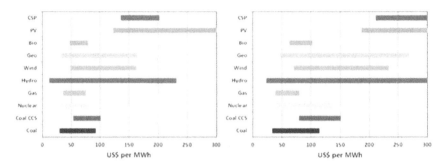

Figure 2: LCOE of different generation options for five per cent discount rate (left) and ten per cent discount rate (right).

Source: IEA/NEA 2010.

85 per cent for coal and natural gas fired generation, respectively.[9] The cost of uranium is about 20 per cent of total nuclear fuel costs, so a doubling of uranium prices increases generating costs about five per cent. A doubling of coal or gas prices raises fossil fuel generating costs between 60 and 80 per cent (Rogner, 2011;[10] IEA/NEA 2010).[11] Nuclear power reduces vulnerability to rising prices and fossil fuel market volatility. Displacement of high-priced fuel imports can also improve balance of payments and reduce drainage of scarce export earnings.

NPPs are among the lowest-cost generators of base load electricity on a grid — usually second only to hydro power (Rogner, 2010).[12] However, low generating costs do not automatically imply low electricity rates or consumer tariffs. In competitive electricity markets, the marginal supplier balancing demand and supply — not

---

[9] In countries with cheap domestic coal (e.g., Australia) or natural gas (e.g. shale gas in the US), the respective fuel cost is lower, so new nuclear build (and even the operation of some existing US plants or their license extension) might not be competitive.

[10] Rogner H-H. (2011) The economics of nuclear power: past, present and future aspects. Chapter 15 in: *Infrastructure and Methodologies for the Justification of Nuclear Power Programmes*. Alonso A. (ed.). Cambridge, UK: Woodhead Publishing Limited.

[11] IEA/NEA (2010). op.cit.

[12] Rogner H-H. (2010). Nuclear Power and Sustainable Development. *Journal of International Affairs*, Fall/Winter 2010, 64(1).

the lowest cost generator — sets the rates, especially during periods of peak demand.

With low greenhouse gas emissions kilowatt per hour (GHG/kWh) nuclear power helps protect against rising costs of future climate change mitigation policies. Any penalty on GHG emissions, (carbon tax or emission cap) would further improve the competitiveness of nuclear power (and renewables) relative to fossil fuel alternatives (see Figure 3 below). Absent a carbon tax, standard coal-fired electricity is the cheapest (red bars) — less than low cost nuclear generation (dashed line) and much less than high cost nuclear (solid line). A 10$/t CO_2 carbon tax would make low cost nuclear power cheaper than coal. High cost nuclear power would require a least a 15$/t CO_2 tax to break even with gas combined cycle technology and an almost 30$/t CO_2 tax to outperform coal. In all cases, nuclear power is competitive versus coal with carbon capture and storage (CCS).

Supply Security

Nuclear energy enhances supply diversification and thus energy security. Uranium reserves and resources are abundant and widely

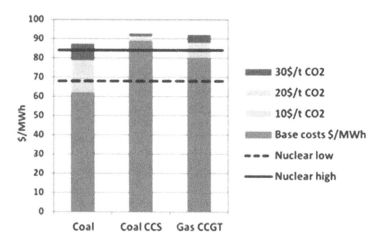

Figure 3:   Impacts of different carbon taxes on the competitiveness of nuclear power.

Source: Author.

distributed geographically. Present uranium resources are sufficient to fuel existing reactors for more than 100 years and for almost 200 years if all conventional uranium occurrences are considered.[13] The high energy density of uranium translates into low fuel volumes and allows for on-site stock-piling of uranium ore for the life of the plant. Long-refuelling cycles (18 to 24 months) plus on-site storage of fuel elements for one refuelling event provide sufficient time to seek alternate suppliers in case the original supplier defaults on contractual arrangements (Rogner and Shihab-Eldin 2013).[14]

Nonetheless, nuclear technology also represents certain supply security risks. Technology recipient countries would, at least initially, be fully dependent on technology and fuel imports from abroad, and vulnerable to politically motivated restrictions such as the 1–2–3 agreement[15] with the US (Rogner and Shihab-Eldin 2013).[16] This agreement roots in weapons proliferation concerns and essentially excludes domestic fuel cycle activities in the partner country and forces abdication of the 'inalienable right' as stated in Article 4 of the Non-Proliferation Treaty (NPT).

Environmental Impacts

Nuclear power is a potent climate mitigation technology. On a life cycle basis, the full technology chain for nuclear energy from

---

[13] Reprocessing of spent fuel and the recycling of unspent uranium and plutonium doubles this figure. Fast breeder reactor technology can further increase uranium utilization over 50-fold IAEA (2012). Climate Change and Nuclear Power 2012. Vienna, Austria: International Atomic Energy Agency.

[14] Rogner H.H and A. Shihab-Eldin (2013). Nuclear Power Option. Chapter 4 in: *Sustainable Energy — Prospects, Challenges, Opportunities*. Gelil I.A., El-Ashry M. and N. Saab (eds.). Beirut, Lebanon: 2013 Report of the Arab Forum for Environment and Development (AFED).

[15] Section 123 of the United States Atomic Energy Act of 1954 governs cooperation in the area of nuclear energy between the US and any other nation. It requires a bilateral agreement between the USA and the recipient country without which US firms may not engage in nuclear technology transfer to that country.

[16] Rogner and Shihab-Eldin (2013). op.cit.

uranium mining to decommissioning being comparable with the best renewable energy chains and at least one order of magnitude lower than fossil fuel chains even with CCS technology (see Figure 4 below). Most NPP GHG gas emissions arise from construction and in the uranium enrichment process. The enrichment industry has been increasingly switching to gaseous centrifuge technology, which requires only about two per cent of the energy input needed for the outdated gaseous diffusion.[17] Clearly, the type of enrichment process considered in any assessment and the extent to which nuclear fuel recycling is accounted for, greatly affect the estimated emissions, resulting in a range of between three and 24 g CO_2-eq. /kWh.

The use of nuclear power also avoids other air and water borne emissions associated with the combustion of fossil fuels chiefly

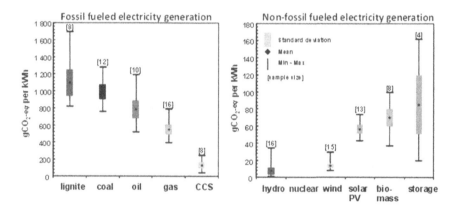

Figure 4: Life cycle GHG emissions of different electricity generating options. Note the one order of magnitude difference in the vertical scales between left and right panels.

Source: Weisser, 2007.[18]

---

[17] Sovacool, B. (2008). Valuing the greenhouse gas emissions from nuclear power: A critical survey, *Energy Policy* **36**, pp. 2, 950–63, reports considerably higher GHG emissions — the result of worst case assumptions regarding components of the nuclear chain such as dated energy-intensive diffusion enrichment technology combined with lowest concentration uranium ore deposits.

[18] Weisser D. (2007). A guide to life-cycle greenhouse gas (GHG) emissions from electric supply technologies. *Energy* **32**(9), pp. 1, 543–1,559.

responsible for poor local air quality and regional acidification with adverse consequences for human health and the environment. Modern fossil-fuelled plants are designed to reduce the emissions of particulate matter, sulphur dioxide and nitrogen oxides to quasi innocuous levels but at a cost.

The operation of nuclear fuel chain facilities releases small quantities of radiation to the environment. These releases are strictly regulated, carefully monitored and controlled to levels less than 0.1 per cent of the public's exposure to radiation from naturally occurring radioactivity arising from radioactive minerals in the ground, from atmospheric radon and from cosmic rays (UNSCEAR 2010).[19] Public exposure to radioactivity from routine nuclear electricity generation is comparable to, or lower than, the radioactivity released by alternative electricity generation or some material-intensive energy efficiency measures. For example, in the US, the annual radiation dose within 50 miles of a coal-fired power station is more than three times that within 50 miles of a nuclear power station (EPA 2012).[20] However, in case of a severe nuclear accident, surface radioactive concentrations in the plant vicinity can be high and can last for decades. In areas further away, agricultural production and fishing may need to be temporarily suspended.

The Three Mile Island accident resulted in the release of minute amounts of radioactive gases with inconsequential health and environmental impacts (UNSCEAR 2011).[21] The Chernobyl and Fukushima Daiichi accidents released large amounts of radioactive

---

[19] UNSCEAR (2010). Atomic Radiation, Sources and Effects of Ionizing Radiation — United Nations Scientific Committee on the Effects of Atomic Radiation. 2008 Report to the General Assembly with Scientific Annexes, Vol. I, New York, USA: United Nations.

[20] EPA (2012). Calculate Your Radiation Dose. US Environmental Protection Agency (EPA), (http://www.epa.gov/radiation/understand/calculate.html; accessed 6 April 2015).

[21] UNSCEAR (2011). Sources and Effects of Ionized Radiation — UNSCEAR 2008 Report to the General Assembly, Volume II, Annexes C, D and E. New York, USA: United Nations.

materials with significant social, economic and environmental consequences, although there have been no radiation related fatalities
in the Fukushima Daiichi accident (UNSCEAR 2012).[22] Latest
analyses estimate the long-term fatalities associated with the
Chernobyl accident at approximately 4,000 to 30,000 late life cancer
deaths.[23] However, non-radiation impacts can be significantly
larger than radiation impacts. Most of the 335,000 evacuees from
the villages around Chernobyl did not return to their original
homes and suffered from depression and stress related difficulties
(Simmons 2012).[24] More fatalities per year are recorded in other
industries like, mining, coal, oil and hydro power than in the
nuclear industry (Burgherr *et al.* 2011).[25]

All electricity generation technology chains generate by-products
and wastes that vary widely in terms of volumes per kWh, toxicity
and longevity. The nuclear chain produces waste of varying levels
of radiotoxicity. Low (LLW) and intermediate level wastes (ILW)
account some 98 per cent of the total volume but only approximately eight per cent of the total radioactivity. LLW (often comparable in radioactivity to waste from other industrial chains) and
ILW arise mainly from routine facility and fuel cycle activities.

---

[22] UNSCEAR (2012). UNSCEAR assessment of the Fukushima-Daiichi accident.
Background Information for journalists. 23 May 2012, (http://www.unis.
unvienna.org/pdf/2012/UNSCEAR_Backgrounder.pdf; accessed?).

[23] Today, there is still uncertainty about future mortalities dues to long latency
periods for many cancers however cancer deaths in Chernobyl affected regions
are expected to be similar to non-Chernobyl controls according to Simmons P.
(2012) The 25th Anniversary of the Chernobyl Accident. Business, Economics and
Public Policy Working Papers, No 2012-1. Australia:School of Business, Economics
and Public Policy Faculty of the Professions University of New England,
Armidale.

[24] Simmons (2012). ibid.

[25] Burgherr P., Eckle P. and S. Hirschberg (2011). Final Report on Severe Accident
Risk including Key Indicators. SECURE — Security of Energy Considering its
Uncertainty, Risk and Economic Considerations. Energy Deliverable No. 5.7.2a.
European Commission Seventh Framework Programme. Villingen, Switzerland:
Paul Scherrer Institute.

Disposal has been practiced safely for decades in many countries using engineered facilities (IAEA 2009).[26]

High level waste (HLW) is the topic of debate. HLW is either spent nuclear fuel or waste separated from reprocessing spent fuel. HLW accounts for two to three per cent of the total nuclear radioactive waste but its radiotoxicity and longevity present particular challenges (IAEA, 2004).[27] Globally, nuclear power plants combined produce approximately 10,000 m³ HLW per year. This would cover the size of a soccer field to a depth of 1.5 metres (GoA 2006).[28]

Although to date no repository accepting civilian nuclear HLW is in operation, the nuclear industry has practiced the safe temporary surface storage of spent fuel for more than half a century with major advances towards the first operating disposal facility. Sweden and Finland already selected disposal sites with full participation of the surrounding communities. Other countries, such as France and Canada, have set out timetables for developing geological disposal facilities. It should be noted that long lived toxicity is not unique to radioactive waste; other forms of hazardous waste, such as mercury, will retain their toxicity forever and thus require indefinite isolation. Photovoltaic cell manufacturing generates some amounts of toxic and hazardous wastes with necessary confinement of thousands of years (ENEF 2010).[29]

---

[26] IAEA (2009). Classification of Radioactive Waste — General Safety Guide. IAEA Safety Standards Series No. GSG-1. Vienna, Austria: International Atomic Energy Agency.

[27] IAEA (2004). Implications of Partitioning and Transmutation in Radioactive Waste Management. Technical Reports Series No. 435. Vienna, Austria: International Atomic Energy Agency.

[28] GoA (2006). Australia's uranium — Greenhouse friendly fuel for an energy hungry world. Government of Australia, House of Representatives Standing Committee on Industry and Resources. Canberra, Australia: Commonwealth of Australia.

[29] ENEF (2010). Strengths — Weaknesses — Opportunities — Threats (SWOT) Analysis. European Nuclear Energy Forum. Working Group Opportunities — Subgroup on Competitiveness of Nuclear Power. Part 1: Strengths & Weaknesses. Luxemburg: European Commission Directorate General for Energy — DG ENER.

Conclusion

The socioeconomic impact of nuclear power is location dependent and cannot be generalised. Over-all, the impacts *per se* are positive. Nuclear power enhances energy security through technology and fuel diversification, and protects against the vagaries of volatile fossil fuel markets. Nuclear power can play an important role in climate mitigation and curbing greenhouse gas emissions, along with the large scale deployment of renewable technologies (once economics and storage accounting for their intermittent nature are demonstrated), and energy efficiency improvements. Macro-economically there are some exceptions. Small open economies are less likely to benefit from nuclear component manufacturing and fuel cycle services than large economies. And in countries with low cost domestic coal or gas, or extensive hydropower, these technologies may well outperform nuclear power.

ABOUT THE CONTRIBUTORS

Prof Dato' Dr Aishah Bidin

Prof Dato' Dr Aishah Bidin is Professor of Corporate and Insolvency Law at the National University of Malaysia. She has been an academic staff of the Faculty of Law of UKM since 1984. Throughout her service in UKM, Prof. Dato' Dr Aishah Bidin used to serve as the Deputy Dean for (Academic, Research and International Relations) and subsequently Dean of the Faculty of Law from 2009–2014. She also served as the Legal Advisor of UKM Holdings (2008–2013), the corporate arm of UKM. In 2013, she was appointed as a Commissioner of the Human Rights Commission of Malaysia (SUHAKAM) for a three-year term where she also served as the Asia Pacific Representative in Global Alliance of National Human Rights Institutions (GANHRI) Working Group on Business and Human Rights (2014–2016). In June 2016, Prof. Dato' Dr Aishah Bidin was re-appointed as a Commissioner for a second term from 2016 until 2019.

Dr Alistair D.B. Cook

Dr Alistair D. B. Cook is Research Fellow at the Centre for Non-Traditional Security (NTS) Studies, S. Rajaratnam School of International Studies (RSIS), Nanyang Technological University (NTU), Singapore. In 2012–2013, he was a visiting Research Fellow at the East Asian Institute of the National University of Singapore.

His research interests are in non-traditional security and human security in the Asia-Pacific including peace and conflict studies, environmental security and climate change, foreign policy and regional cooperation, and domestic politics in Myanmar. He graduated with a PhD from the University of Melbourne, Australia; Masters from Purdue University, USA; and M.A. (Hons) from St. Andrews University, Scotland. He has taught at Purdue University, University of Melbourne, Deakin University, Nanyang Technological University and Australian National University.

Dr Eulalia Han

Dr Eulalia Han was a Research Fellow, Energy Security Division, Energy Studies Institute, National University of Singapore. She holds a PhD in International Relations from Griffith University and a BA in Political Science and International Relations (Class 1 Honours) from the University of Queensland. She has lectured, tutored and designed course materials for undergraduate students and also worked as a Policy Officer with the Department of Emergency Services in Queensland, Australia. Her publications include journal articles in the Australian Journal of Political Science and Media International Australia.

Professor Dr Hans-Holger Rogner

For most of his career, Dr Rogner has been engaged in comprehensive energy system analysis, energy modelling and integrated resource planning. Until his retirement from the International Atomic Energy Agency (IAEA) in 2012, he directed the International Atomic Energy Agency's programme on Capacity Building and Nuclear Knowledge Maintenance for Sustainable Energy Development. Dr. Rogner holds positions of Affiliate Professor, Royal Institute of Technology (KTH), Stockholm, Sweden and Senior Research Scholar at the International Institute for Applied Systems Analysis (IIASA) in Laxenburg, Austria. Dr. H-Holger

Rogner holds an MSc in Industrial Engineering (1975) and a PhD in Energy Economics (1981).

Dr Eur Ing John A Bauly

Dr Eur Ing John Bauly has over 25 years of commercially successful "hands on" involvement in business management, product development and R&D. He has extensive experience in nuclear and fossil power plant, telecommunications, electronics, aircraft and automotive products, — from working with those industries in Europe, North America, Singapore, Japan and Indonesia. He now works in the S.E Asia and Europe as a consultant, energy analyst, and corporate educator. In recent months he has presented 2-3 day short course coaching seminars in Singapore, Shanghai, Sydney, Bangkok, Malaysia, and London. He also lectured and researched at the National University of Singapore for a few years and still maintains close links with them — particularly for energy studies on the modelling of trajectories to lower carbon in ASEAN and Europe, and nuclear power governance and risks. He has published widely on energy subjects, NPD, and R&D. See also https://www.linkedin.com/in/john-bauly-1328519?

Mr Nicholas Fang

Mr Nicholas Fang is Singapore Press Holdings scholar with a Masters Degree from Oxford University. He worked at The Straits Times national newspaper for nine years, rising to the post of Senior Correspondent before he moved to Channel NewsAsia at MediaCorp Singapore as business desk editor and presenter, taking over anchoring of major news bulletins, including Budget specials. He is also the executive director of Singapore's oldest political think tank, the Singapore Institute of International Affairs. He founded Black Dot sports consultancy in 2012 to drive growth of the sports sector in Singapore. Appointed as a Nominated Member of Parliament in Singapore in 2012 focusing on international affairs, sports, defence and media. Nicholas rejoined Channel NewsAsia in 2013 as associate editor overseeing coverage of Singapore news.

Mr Nur Azha Putra Bin Abdul Azim

Nur Azha Putra is a Research Associate with the Energy Security Division at the Energy Studies Institute, National University of Singapore. Prior to his appointment, he was an Associate Research Fellow at the Centre for Non-Traditional Security Studies at the S. Rajaratnam School of International Studies (RSIS), Nanyang Technological University (NTU), Singapore, where he was the lead researcher on the Energy Security and the NTS-Plus programmes. Before joining RSIS, he was a Research Officer at the Centre for Research on Islamic and Malay Affairs (RIMA), a subsidiary and research wing of the Association of Muslim Professionals, Singapore. Before moving into the research field, he was a journalist with Singapore Press Holdings' national Malay newspaper, Berita Harian and an Information Systems Officer at a regional brokerage firm. Azha graduated from NTU with a Master of Science in International Political Economy. He also holds a Bachelor of Information Technology from Central Queensland University, Australia.

Ms Sofiah Jamil

Sofiah Jamil commenced her PhD in March 2012 and is a recipient of the ANU University Research Scholarship. She was conferred an MSc (International Relations) from Nanyang Technological University, Singapore in 2010, during which she wrote a dissertation on "Democracies and Effective Climate Change Mitigation: The Case of Indonesia". She received her Bachelor of Arts (2nd Upper Honours) in Political Science and International Relations from the University of Western Australia in 2006. Her honours thesis is entitled "Oh I See!" — The Organisation of Islamic Conference's Enlightened Moderation Agenda. Prior to pursuing her PhD, Sofiah was an Associate Research Fellow at the RSIS Centre for Non-Traditional Security (NTS) Studies, where she currently holds an Adjunct Research Associate position. She also serves on the Board of Management of the Young Association of Muslim Professionals in Singapore.

Madam Ton Nu Thi Ninh

Ambassador Ton-nu-thi Ninh is a member of Vietnam's law-making body, the National Assembly, representing the southern coastal province of Ba Ria — Vung Tau. In her position as Vice-Chair of the National Assembly Foreign Affairs Committee, her mission has been to develop and enhance Vietnam's relations with the countries of North America (particularly, the United States) and Western Europe. She travels frequently to the United States and Europe and regularly interacts with senior government and business leaders both abroad and in Vietnam. She has also represented Vietnam in international conferences among world leaders to discuss issues with global implications. She is widely recognized as an effective spokesperson for Vietnam.